Use R!

This series of inexpensive and focused books on R is aimed at practitioners. Books can discuss the use of R in a particular subject area (e.g., epidemiology, econometrics, psychometrics) or as it relates to statistical topics (e.g., missing data, longitudinal data). In most cases, books combine LaTeX and R so that the code for figures and tables can be put on a website. Authors should assume a background as supplied by Dalgaard's Introductory Statistics with R or other introductory books so that each book does not repeat basic material.

How to Submit Your Proposal

Book proposals and manuscripts should be submitted to one of the publishing editors in your region per email – for the list of statistics editors by their location please see https://www.springer.com/gp/statistics/contact-us. All submissions should include a completed Book Proposal Form.

***

For general and technical questions regarding the series and the submission process please contact Faith Su (faith.su@springer.com) or Veronika Rosteck (veronika.rosteck@springer.com).

Nabeel Siddiqui

Cultural Analytics in R: A Tidy Approach

Springer

Nabeel Siddiqui
Susquehanna University
Selinsgrove, PA, USA

ISSN 2197-5736 ISSN 2197-5744 (electronic)
Use R!
ISBN 978-3-031-96617-0 ISBN 978-3-031-96618-7 (eBook)
https://doi.org/10.1007/978-3-031-96618-7

This Springer imprint is published by the registered company Springer Nature Switzerland AG
The registered company address is: Gewerbestrasse 11, 6330 Cham, Switzerland

For Azmeh, Zayaan, and Iman

Declarations

Competing Interests The authors have no competing interests to declare that are relevant to the content of this manuscript.

Contents

Chapter 1
Introduction

Abstract This chapter lays the groundwork for a comprehensive exploration of cultural analytics, an emerging discipline at the intersection of data science and humanities. It begins with a historical overview, identifying the transition from a culture of scarcity to a culture of abundance in the digital age and the challenges and opportunities this presents for humanities research. It advocates for a paradigm shift in humanities scholarship, emphasizing the need for new methodologies to analyze and interpret the vast digital archives now accessible to researchers.

It introduces the concept of "cultural analytics" as a field that utilizes computational and visualization techniques to uncover insights from massive cultural datasets that transcend the traditional divide between quantitative and qualitative methods. It goes on to explore the specifics of using the R programming language for cultural analysis, highlighting its potential due to its powerful package ecosystem and its ability to handle a variety of data types. The concept of tidy data is presented as a necessary standard for efficient data analysis in cultural analytics, with an emphasis on the "tidyverse" suite of packages that simplifies data manipulation for humanities scholars.

Keywords Cultural analytics · Digital humanities · Big data · Methodological shift · R programming · Tidy data · Tidyverse · Quantitative analysis · Qualitative analysis · Data visualization · Computational methods · Interdisciplinary research · Data science · Humanities S-scholarship

1.1 Cultural Abundance and the Rise of Cultural Analytics

This book is the result of almost 15 years spent thinking about the role of data in the humanities. As an undergraduate history major in the mid-2000s, I was one of the first to be able to conduct the majority of my research online. Resources were being increasingly digitized, and scholars across the globe could now access a treasure trove of books, articles, manuscripts, and other primary sources. For some scholars, growing digitization meant more efficient access to existing knowledge, but for others, it signaled the dawn of a new standard for the historical profession.

N. Siddiqui, *Cultural Analytics in R: A Tidy Approach*, Use R!,
https://doi.org/10.1007/978-3-031-96618-7_1

No longer were historians limited by their geographical location or by the physical constraints of their libraries. Instead, they could now analyze massive amounts of data to uncover patterns, trends, and insights that were previously invisible.

One pioneer in understanding the implications of this shift toward the digital was historian Roy Rosenzweig. In a 2003 article for the *American Historical Review*, he implored historians to confront the digital revolution head-on. As he contended, historians were now "facing a fundamental paradigm shift from a culture of scarcity to a culture of abundance" (Rosenzweig 2003).

While a growing number of historians have teamed with archivists to assist in the digitization of resources, there has been limited response to Rosenzweig's additional call for deeper exploration of the epistemological and methodological shifts that digital abundance may bring about. Rosenzweig recognized that, although the Internet presented opportunities, it also opened up a Pandora's box of complexities. Given the quantity of the material, "reading around" to generalize large trends, which had been the traditional approach to historical research, was no longer feasible. Presumably, even marginal viewpoints could now be found in abundance online, raising concerns about how much material was needed to be representative of a given phenomenon. Historian Ian Milligan makes a similar point: "At the scale we are now working within born-digital data, the mere fact of something existing—or even hundreds of something existing—may not signify something significant" (Milligan 2016, 57). Historians need to develop new frameworks and methodologies to analyze and interpret the expansive digital archives. "Graduate programs will probably have to teach such social-scientific and quantitative methods as well as such other skills as 'digital archaeology' (the ability to 'read' arcane computer formats), 'digital diplomatics' (the modern version of the old science of authenticating documents), and data mining (the ability to find the historical needle in the digital haystack)," contended Rosenzweig (Rosenzweig 2003).

A focus on the debates surrounding historical methodologies may seem specific to that field, but the questions they raise apply to the study of culture more broadly. Digital artifacts are now all around us, and fields as diverse as English, film studies, musicology, media studies, communications, and cultural studies are grappling with similar issues surrounding questions of scale, preservation, and interpretation in an era of growing information. New inquiries have emerged in light of these developments. How does one sift through such voluminous data to find what is truly significant? What tools and techniques can be applied to manage and interpret this information? What role does data visualization play in understanding and communicating insights from extensive datasets? Are there specific tools or software that can assist in data management and analysis? These questions pose considerable challenges, but they also afford extraordinary opportunities for exciting innovation.

To make sense of the changes brought on by the digital revolution in the humanities, some scholars have posited the emergence of "cultural analytics" as a stand-alone discipline. Coined by Lev Manovich, cultural analytics refers to "the analysis of massive cultural data sets and flows using computational and visualization techniques" (Software Studies Initiative 2015). For him and others in the field, cultural analytics does not simply involve applying quantitative analysis to cultural data. Instead, it represents a significant epistemological shift, where new insights are gleaned from patterns and metadata that are only discernible through computational means.

Manovich contrasts cultural analytics with two related fields that it often intersects with: digital humanities and computational social sciences. As he notes, "In keeping with the historical orientation of the humanities, digital humanities scholars use computers to analyze mostly historical artifacts created by professionals" (Manovich 2016). On the other hand, computational social scientists are more concerned with contemporary phenomena, using digital tools to assess large-scale social patterns and behaviors. Cultural analytics intersects and enriches these areas by focusing on both contemporary and historical cultural data. Additionally, it is not limited to professional productions but is inclusive of everyday artifacts of social media activity. For instance, one cultural analytics scholar may be interested in analyzing user-generated content on platforms like YouTube or TikTok to understand the impact of grassroots cultural movements, while another may want to explore nineteenth-century Victorian novels.

Another defining feature of cultural analytics is its disavowal of a strict divide between quantitative and qualitative methods and its advocacy for an approach that values the integration of both. In other words, it is "interested in combining both in the study of cultures—focusing on the particular, interpretation, and the past from the humanities, while centering on the general, formal models, and predicting the future from the sciences" (Manovich 2016). By analyzing large cultural datasets through the tools of data science, such as machine learning and network analysis, scholars can reveal patterns and trends that allow us to move from the general to the specific. Perhaps the trend shows us the pervasiveness of a particular narrative structure in contemporary fan fiction, or it enlightens us to the spread of certain Internet memes. Thus, while traditional humanities scholarship may focus on close readings and meticulous case studies, cultural analytics can offer broad overviews that contextualize these specific studies within larger cultural movements.

Although cultural analytics holds promise for unraveling complex phenomena through digital means, it requires humanities scholars to embrace programming and coding. Embracing these tools not only expands the methodological repertoire of humanities scholars but also enables them to engage in interdisciplinary collaborations and contribute to the evolving field of cultural analytics. Yet, the myriad of programming languages and environments can present a steep learning curve for those unfamiliar with them. The primary objective of this book is to bridge the divide between traditional humanities education and the emerging needs of the field. It achieves this by offering a comprehensive introduction to computational methods customized for humanists using the versatile programming language R.

1.2 Why R for Cultural Data?

As cultural analysts deal with more data, we will need tools that are powerful enough to process and analyze vast and complex datasets efficiently. The choice of R for cultural analysis, however, may initially seem unconventional given its strong emphasis on statistics. As Norman Matloff notes, "R is written by statisticians, for statisticians," and this is reflected in its history (Norman 2014). Developed in the early 1990s at the University of Auckland by Ross Ihaka and Robert Gentleman, R emerged as a dialect of the programming language S. Ihaka and Gentleman were seeking out a piece of software that they could use in their Macintosh teaching lab oriented for statistics but found no suitable options. Consequently, they set out to create a language that would meet their needs and, in doing so, laid the foundation for one of the most versatile and widely used statistical programming environments for academic statisticians (Ihaka 1998).

From the beginning, R developed a large culture that encouraged the sharing of code among its users. In March 1997, Hornik and Leisch, of the Vienna University of Economics and Business, built the Comprehensive R Archive Network (CRAN), which serves as a repository of thousands of packages that extend the base R system (Thieme 2018). This system resulted in a paradigm shift where statisticians and data analysts could easily access a wealth of tools tailored to their specific needs. In the field of statistics and other quantitatively heavy fields, for example, it is still common to publish new methods in journal articles that are accompanied by R packages. These packages serve as a mechanism not just for dissemination but also for verifying computational claims, thereby fostering replicability and rigor in scholarly communication (Xie 2018). Notably, the wide range of libraries developed for the language enables new users, including those in the humanities, to tailor its capabilities to their specific research questions.

Different subfields in the humanities, especially the digital humanities and cultural analytics, have increasingly recognized R's potential and harnessed its capabilities for diverse research endeavors. As Arnold et al. note, the "R language has acquired considerable support within the Digital Humanities" (Arnold and Tilton 2019). They highlight the growing amount of training at various annual conferences and workshops, such as HILT and DHSI, dedicated to R along with a series of books in this series, such as *Humanities Data in R* (Arnold and Tilton 2015), *Text Analysis with R for Students of Literature* (Jockers and Thalken 2020), and *Corpus Linguistics and Statistics with R* (Desagulier 2017) as evidence of this growing acceptance.

A chief catalyst for R's growth in the humanities is its ability to accommodate the diverse range of data encountered by cultural analytic researchers. Scholars working in this field often need to analyze text, numbers, images, or geospatial information within a single project. According to Arnold and Tilton, "Using a general-purpose statistical programming language greatly increases the available set of methodological approaches to studying humanities data" (Arnold and Tilton 2019). However, each type of data may have unique underlying mechanisms for processing and analysis. This is where R's powerful package ecosystem comes into play. It has coalesced around a standardized data structure called "tidy data," which simplifies the process of working with diverse datasets by imposing a consistent structure that allows researchers to easily manipulate and analyze their data. This standardization provides researchers with increased efficiency and confidence so that they can focus on exploring and interpreting their data, rather than getting lost in the intricacies of data management.

1.3 Why Tidy Data?

In the evolving terrain of cultural analytics, the meticulous study of cultural patterns, trends, and phenomena relies heavily on the insightful interpretation of complex datasets. As Christof Schöch notes, we can think of the humanities as having two types of data: big data and smart data (Schöch 2013). The former refers to the volume of the data itself and is, in many cases, in the eye of the beholder. For instance, a humanities scholar may see a Victorian novel as a complex big data problem, needing text mining techniques to unearth underlying themes and narrative structures. On the other hand, a biostatistician might only consider data "big" if it involves terabytes of genomic information. What is important about this type of data is not necessarily its volume but the fact that it is usually unstructured and "messy." In contrast, smart data is structured or semi-structured. It contains metadata and annotations that aid in the interpretation of the underlying content, making the data more navigable and useful for specific inquiries.

As Schöch argues, big data and smart data are often in opposition to one another. While big data is usually messy, smart data is usually small. He sees bridging the gap between the two as one of the chief challenges for humanities scholars moving forward and posits the use of automatic annotation and crowdsourcing as one way to accomplish this. While these techniques are useful, they often presuppose an existing framework or schema about how data should be organized. To date, though, humanities scholars have paid little attention to data organization. Fortunately, they can tap into the extensive body of research and knowledge available in fields such as relational database management, thereby providing them with a valuable resource that they can leverage and integrate into their own work.

Since the 1970s, experts in relational database management have developed a set of principles known as normalization to facilitate data structuring. These principles, commonly referred to as normal forms, provide guidelines for designing efficient and flexible databases and are based on the idea of eliminating data duplication and maintaining consistency (Codd 1990). There are several normal forms, each with its own set of rules and requirements, with the most common being the first, second, and third normal forms (1NF, 2NF, and 3NF). In the context of large databases, understanding and adhering to these principles is essential to ensuring that queries yield accurate results and the system performs optimally. Yet, most humanities research does not rely on such vast amounts of data, thereby posing the question of how these traditional database management principles can be effectively applied within the humanities sphere.

The concept of "tidy data," as coined by Hadley Wickham, provides a bridge between the principles of relational database management and the needs of humanities researchers. According to Wickham, tidy data broadly refers to "Codd's 3rd normal form, but with the constraints framed in statistical language, and the focus put on a single dataset rather than the many connected datasets common in relational databases" (Wickham 2014). The key principles of tidy data are as follows (Wickham et al. 2023):

1. Each variable forms a column.
2. Each observation forms a row.
3. Each value is a cell; each cell is a single value.

For now, these concepts will largely seem abstract—don't worry! We will look at what they mean in more depth in Chap. 2—but applying these principles can significantly improve the workflow for humanities researchers. Not only does it make it easier to understand and manipulate data, but it also ensures consistency throughout the research process. This means fewer errors during analysis and a more straightforward approach to data visualization and interpretation.

Most importantly, the concept of "tidy data" has given rise to an ecosystem of packages collectively known as the "tidyverse," which empowers scholars in cultural analytics to efficiently clean, organize, and transform their data according to the principles of tidy data (Wickham et al. 2019). The "tidyverse" packages are intuitive and user-friendly, offering commands that correspond more closely to natural language. This is especially beneficial for those who may not have a strong programming background. Perhaps most importantly, it allows humanities scholars to work with a variety of data types without having to learn a multitude of different system-specific commands. This interoperability is vital for projects that cross traditional disciplinary boundaries and call for a fluid exchange of methodologies.

1.4 Who Is This Book for?

Despite the growing amount of cultural data available, training in the humanities remains tied to "traditional" approaches that sometimes overlook the analytic possibilities offered by digital methods. Few humanities scholars today have training in quantitative methods or possess the coding skills that are essential to their peers in computer science or data science. In fact, for many humanities scholars, the very notion of "data" can seem alien and overly empirical, an imposition of scientific methodology on the study of human culture and history that may not always seem appropriate. As Nick Redfern argues, humanities scholars see their disciplines as bound by an epistemological foundation that relies on qualitative analysis, which creates a resistance to approaches that may be perceived as reductionist. Thus:

> "Because the methodologies used by digital humanists come from the sciences... the knowledge they produce is the product of the scientific method and is, therefore, scientific knowledge. By definition, the knowledge produced by the digital humanities cannot be a part of humanist knowledge about culture because it is produced by the wrong way of knowing." (Redfern 2023)

This book does not aim to engage in the theoretical debates surrounding the use of quantitative methods in the humanities. Instead, it assumes that the reader is a humanities scholar who is already interested in exploring the potential of data analysis for their research. Even for those who are open to integrating digital methods, obstacles can arise due to the lack of resources and learning tools tailored specifically to the humanities. As a graduate student, I only developed a foundation in quantitative methods after seeking out courses outside of my department. This book aims to fill that gap by providing a comprehensive, practical guide to digital humanities methods for students and scholars who may not have a strong foundation in data science and visualization.

This book assumes little familiarity with data science or programming and none in R itself. In fact, for those coming with a background in other programming languages, some of the concepts in R may seem counterintuitive at first due to their unique syntax. I have purposefully avoided as much math as possible and explained concepts through examples from the humanities whenever they are introduced. In this way, readers can see how these data science principles apply directly to their own fields and research questions. By grounding the discussion in familiar contexts, I hope to demystify the process and encourage a more intuitive understanding of the material. My goal is that those interested in understanding the background of the methodologies employed will find in this book both a guide and an encouragement to explore further.

That said, I will assume that you can download R and install the necessary packages on your computer. In the case that you encounter difficulties, resources are widely available online. For art historians, literary theorists, musicologists, and other cultural researchers, the thought of installing programming software and packages may seem daunting. However, the process has been significantly simplified over recent years, making it achievable even for those who may not consider themselves technically adept.

Above all, this book is for the inquisitive humanities scholar looking to bridge the gap between traditional qualitative methods and the quantitative tools that have become increasingly essential in the digital age. For the experienced computer science professional or the data analyst, many of the procedures and analyses discussed may seem elementary or familiar. However, it is precisely this foundational knowledge that can demystify data analysis for those who are newer to the field. By starting with the basics, the book scaffolds the learning process, so readers are not overwhelmed by more complex concepts before they're ready. By the end of this book, humanities scholars will have a solid grounding in digital methods, allowing them to conduct their research with an enriched toolkit that complements their existing expertise.

1.5 Downloading Datasets

The datasets used in this book are available in the following link:

GitHub Repository: All datasets are available at https://github.com/nabsiddiqui/cultural_analytics_in_r/. You can either:

- Download the complete `data.zip` file.
- Access individual chapter datasets from the "data" folder. If doing so, make sure to create a "data" folder in your working directory and place the downloaded files there to match the code examples in the book.

Please ensure to place the downloaded files in a "data" folder within your working directory to match the code examples in the book.

1.6 An Overview of the Book's Structure

In this book, our primary focus will be on exploring the capabilities of R for data analysis in the humanities rather than delving deeply into specific research questions. While research questions are important and ultimately drive the analysis process, our goal here is to provide a foundation and understanding of the tools and techniques that R offers. Throughout the book, we will utilize a diverse range of datasets that showcase the broad applicability of the programming language. While certain chapters in this book can be read independently, it is recommended to read them sequentially to build a solid foundation of R skills and concepts. Each chapter builds on the previous ones, gradually introducing new techniques and tools.

Chapter 1 serves as an essential foundation for those embarking on the journey of learning R. This chapter delves into the basics of base R, the backbone of the R programming language, establishing a solid understanding that is critical before exploring higher-level packages, such as those in the tidyverse ecosystem. By tackling core programming principles, readers will learn how to perform basic mathematical operations, manipulate data structures, and leverage the language's built-in functions for efficient data analysis. The chapter also offers a step-by-step guide through the nuances of vectors, variable assignment, and control structures like loops and conditional statements, emphasizing hands-on practice. To solidify the concepts in the chapter, it showcases how to apply them to a real-world dataset: an extensive compilation of over 120 years of Olympic athletes, which provides a unique lens into historical trends and patterns in the world of sports.

Chapter 2
Understanding Base R

Abstract This chapter provides a comprehensive introduction to R programming. Readers gain a solid understanding of the fundamental concepts of R, ranging from basic mathematical operations and variable assignments to more complex data structures like vectors, data frames, and lists. Through hands-on exploration of a real-world dataset covering 120 years of Olympic athletes, the chapter demonstrates how base R functions can be applied to clean, subset, and analyze data to uncover insights into trends such as health, popularity, nationalism, and gender dynamics. Control structures such as conditional statements and loops are explained, with an emphasis on their practical application in R scripts. Finally, to visually represent data, the chapter covers basic plotting techniques using base R graphics, setting the stage for more advanced visualization tools available in the tidyverse.

Keywords Base R · Data analysis · Vectors · Data frames · Lists · Control structures · Apply functions · Data visualization · Cultural analytics

2.1 Introduction

R has a significant amount of peculiarities. Because of this, many individuals have now started working directly with the "tidyverse" series of packages. While this book makes the case that the tidy framework is particularly beneficial for scholars in cultural analytics, this should not be taken as a dismissal of the importance of understanding base R. In many cases, base R solutions are not only faster but more memory-efficient than their tidyverse equivalents, and there will be instances when you'll need to work with base R functions or packages that follow different conventions. Knowing how to do this will allow you to adapt to these situations and make your work more versatile.

Starting with base R also allows us to avoid jumping directly into more advanced programming concepts that lie at the "tidyverse's" core, such as functional programming, which often eschews concepts like loops and mutable variables in favor of higher-order functions and piped operations. Skeptics of the "tidyverse" have posited that these concepts result in code that is not only harder for beginners

N. Siddiqui, *Cultural Analytics in R: A Tidy Approach*, Use R!,
https://doi.org/10.1007/978-3-031-96618-7_2

to understand but also results in more complicated syntax and potentially slower execution times (Matloff 2023).

Throughout this chapter, we will explore how R handles basic data types and structures such as vectors, matrices, and data frames. We will discuss control flows in R, including conditional statements and loops, which are fundamental for writing more complex scripts. We will also consider how to write functions in R, which is crucial for anyone looking to write clear, concise, and reusable code.

Most importantly, we will look at a real-world dataset covering 120 years of Olympic athletes (Griffin 2018). This dataset allows us to explore various aspects of the Olympics, including health, popularity, nationalism, and gender dynamics over time. The dataset contains information on more than 200,000 athletes, with each athlete's record including details such as their name, sex, age, height, weight, team, nationality, and the year and season of the Olympic games they competed in. We will explore questions such as which sports show the most significant gender disparities and how the age distribution of athletes has changed over the years. We will apply essential base R functions to clean and manage the data, creating meaningful subsets and performing fundamental analyses.

This hands-on approach will help reinforce the programming concepts discussed, allowing readers to see the immediate applicability of these techniques in a real-world context. By the end of the chapter, you will have a better appreciation for the utility and power of base R in the data analysis process. Moreover, you will be equipped with the knowledge to tackle your analytical challenges using the principles you have learned thus far. Don't worry if you feel overwhelmed at first. This chapter covers a lot of ground, but remember that practice is key when learning any programming language. As you work through the examples and exercises provided, your understanding will grow, and you'll become more comfortable with the language's syntax and concepts.

2.2 Basic Mathematical Operations and Variable Assignments

In R, you can perform basic mathematical operations directly in the console. The standard operators for addition, subtraction, multiplication, and division are `+`, `-`, `*`, and `/`, respectively. Besides these, R allows exponentiation using `^` for power operations. The modulus operator `%%` returns the remainder after division, and the integer division operator `%/%` gives the quotient.

When performing operations, the order of precedence follows the standard mathematical conventions, and you can use parentheses `()` to control the sequence of the operations. This makes tackling complex expressions more manageable.

Two things are worth noting. One, R will print the result to the console by default. However, to save a result for later use, you would need to assign it to a variable, which we will look at briefly. Two, you can add comments by beginning any line with (#). These comments are not executed but are immensely helpful in documenting your code to make it more readable to you and others:

```
# Addition
2 + 5
```

```
[1] 7
```

```
# Subtraction
7 - 3
```

```
[1] 4
```

```
# Multiplication
4 * 6
```

```
[1] 24
```

```
# Division
10 / 3
```

```
[1] 3.333333
```

```
# Exponentiation
3^2
```

```
[1] 9
```

```
# Integer division
12 %/% 5
```

```
[1] 2
```

```
# Outputs the result of the operation
(2 + 3) * (5 - 2)
```

```
[1] 15
```

As you can see, each of these operations is quite straightforward. However, the real power of R comes from its ability to operate on other data structures. In R, just like other programming languages, you can use variables to store values to manipulate and analyze later in your code. To assign a value to a variable, we use the assignment operator, which is an arrow made up of a less-than symbol and a hyphen (<-). Although you can also use the equals sign (=) for assignment, it is recommended to stick to the (<-) operator, as it is more common and less prone to errors in certain cases:

```
# Assign the value 5 to the variable x
x <- 5

# Assign the value 10 to the variable y using the equal sign
y = 10

# Assign the result of the addition of x and y to the variable sum_xy
sum_xy <- x + y
```

Keep in mind that variable names are case-sensitive, so “a” and “A” would be considered separate variables. Try to use descriptive names that are easy to understand when naming your variables. In the past, it was common for many R programmers to use a dot in variable name assignments, such as “sum.xy.” Because this convention is used in object-oriented programming languages to access the properties of an object, it has grown out of favor (Wickham n.d.). Avoid this to make your code easier to read, both for yourself and others.

2.3 Vectors in R

R has several data types. Understanding these is essential to avoiding potential errors. The most fundamental data type in R is the vector. A vector is a one-dimensional collection of values and is similar to arrays in other programming languages. There are two main types of vectors: atomic and lists. For now, we will stick with atomic vectors. Atomic vectors come in four main types. Vectors consist of elements of the same data type, such as integers, numerical values, or characters. Later, we will also look at lists, which allow us to combine different data types.

2.3.1 Scalars

The simplest atomic vector is a scalar. Although R does not have a native scalar type, a scalar in R is simply a vector with a single element. You can think of them broadly as numbers. Thus, you have already been working with vectors when you assigned values to variables. Notably, scalars are treated the same as vectors when it comes to operations and functions. This design choice allows R to treat scalars and vectors consistently, and you can perform mathematical operations, apply functions, and manipulate scalars just like you would with vectors.

Scalars can hold a myriad of data types, including logical (TRUE or FALSE), numerical, and character, integer, or complex numbers:

```
# Integer scalar
integer_scalar <- 3

# Numeric scalar (floating-point number)
numeric_scalar <- 3.14

# Character scalar
character_scalar <- "Hello"

# Logical scalar (TRUE or FALSE) # See the output of our scalar
↪  assignments
logical_scalar <- TRUE
```

We can look at the values for each of these vectors by simply calling the variable name:

```
# Print the integer scalar
integer_scalar
```

```
[1] 3
```

```
# Print the numeric scalar
numeric_scalar
```

```
[1] 3.14
```

```
# Print the character scalar
character_scalar
```

```
[1] "Hello"
```

```
# Print the logical scalar
logical_scalar
```

```
[1] TRUE
```

2.3.2 Vectors with More than One Value

You can also use atomic vectors to hold multiple values of the same data type. To do this, we can use the c() function, which stands for concatenate. To use the c()

function, you simply provide the values or vectors inside the parentheses, separated by commas. Here's an example:

```
# Integer vector with more than one value
integer_vector <- c(1, 2, 3, 4, 5)

# Numeric vector with more than one value
numerical_vector <- c(3.14, 2.71, 1.62)

# Character vector with more than one value
character_vector <- c("Hello", "World", "R")

# Logical vector with more than one value
logical_vector <- c(TRUE, FALSE, FALSE)
```

Once you have created a vector of multiple values, you can perform operations, such as accessing individual elements, computing their length, or manipulating values within the vector. In R, indices start at 1, which means that the first element of a vector has an index of 1, the second element has an index of 2, and so on. This is different from some other programming languages, like Python or Java, where indices start at 0.

To access an individual element of a vector in R, you can use square brackets (`[]`) and provide the index of the element you want to access inside the brackets. For example, if you have a vector called "my_vector" with elements [10, 20, 30, 40, 50], you can access the first element (10) using `my_vector[1]`, the second element (20) using `my_vector[2]`, and so on:

```
# Access the first element of the integer_vector
integer_vector[1]
```

```
[1] 1
```

```
# Access the second element of the numeric_vector
numerical_vector[2]
```

```
[1] 2.71
```

```
# Access the third element of the character_vector
character_vector[3]
```

```
[1] "R"
```

```
# Access the first element of the logical_vector
logical_vector[1]
```

```
[1] TRUE
```

It's important to note that when accessing elements, the index should be a valid integer within the range of the vector's length. If you try to access an index that is out of range, R will return an error. If you are unsure, you can get the length of the vector by calling the `length()` function. We will look at how functions work briefly, but for now, you can think of them as pre-written code blocks that you can call upon to perform specific tasks. They normally take an argument to specify what data to operate on and sometimes additional parameters. To use the `length()` function, we place the name of the vector we place parentheses around the vector's name. Here's an example using our previously created vectors:

```
# Get the length of the integer_vector
length(integer_vector)
```

```
[1] 5
```

```
# Get the length of the numeric_vector
length(numerical_vector)
```

```
[1] 3
```

```
# Get the length of the character_vector
length(character_vector)
```

```
[1] 3
```

```
# Get the length of the logical_vector
length(logical_vector)
```

```
[1] 3
```

We also access multiple elements of a vector by providing a vector of indices using the `c()` function. This allows us to retrieve a subset of elements from the original vector. The resulting vector will contain the elements at the specified indices, in the same order as they appear in the index vector. If an invalid index is provided, R will return an error. Additionally, if there are duplicate indices in the index vector, the resulting subset will contain duplicate elements:

```
# Access the 1st, 3rd, and 5th elements of integer_vector (1, 3, and 5)
integer_vector[c(1, 3, 5)]
```

```
[1] 1 3 5
```

```
# Access the 1st and 2nd elements of numeric_vector
numerical_vector[c(1, 2)]
```

```
[1] 3.14 2.71
```

2.3.3 Modifying and Manipulating Vectors

You can modify values within a vector by using indexing and assignment. To update a specific element, you can access it using square brackets and assign a new value to it. To add a new element to the end of a vector, you can use the concatenate function `c()` or access an index that does not currently exist and assign a value to it. Let's look at an example:

```
# Replace the first element with the value 10
integer_vector[1] <- 10

# Add a new element with the value 6 to the end of integer_vector
integer_vector[6] <- 6

# Different way to add a the value 7 to the end of integer_vector
integer_vector <- c(integer_vector, 7)

# Examine the integer vector after modifications
integer_vector
```

```
[1] 10  2  3  4  5  6  7
```

Besides updating a single element or adding new ones, you can remove elements from a vector using the negative index inside the square brackets (`[]`). For instance:

```
# Remove the first element of integer_vector
integer_vector[-1]
```

```
[1] 2 3 4 5 6 7
```

```
# Examine the integer vector after modifications
integer_vector
```

```
[1] 10  2  3  4  5  6  7
```

When performing arithmetic operations on vectors, R will apply the operation element-wise, meaning it will operate on corresponding elements in each vector. This is useful for performing a broad range of mathematical computations efficiently. R will automatically print the output unless you assign it to a new variable. Here are some examples of vector arithmetic operations in R:

```
# Create a vector containing the elements 1, 2, and 3
vector_a <- c(1, 2, 3)

# Create a vector containing the elements 4, 5, and 6
vector_b <- c(4, 5, 6)

# Add the vectors element-wise
vector_a + vector_b
```

```
[1] 5 7 9
```

```
# Multiply the vectors element-wise
vector_a * vector_b
```

```
[1]  4 10 18
```

```
# Divide the vectors element-wise
vector_a / vector_b
```

```
[1] 0.25 0.40 0.50
```

When working with vectors of different lengths, R will perform recycling, meaning it will repeat the shorter vector until it is long enough to match the length of the longer vector. This ensures that the operation can be performed element-wise between the vectors. While this can be confusing at first, it can be a powerful and efficient method for performing calculations on large datasets, especially when working with repetitive patterns or performing calculations involving repeated values. Nonetheless, it's important to be mindful of the lengths of your vectors and the resulting output to ensure that your calculations are accurate and meaningful:

```
# Create a vector called with values 1, 2, 3
vector_x <- c(1, 2, 3)

# Create a vector called vector_y with values 10, 20
vector_y <- c(10, 20)

# Recycle shorter vector to match length of longer vector
vector_x + vector_y
```

```
[1] 11 22 13
```

2.3.4 Logical Operations of Vectors

Logical operations can also be applied to vectors. These operations include comparisons such as equality (==), inequality (!=), greater than (>), less than (<), greater than or equal to (>=), and less than or equal to (<=). When performing logical operations on vectors, R will return a logical vector with the same length as the input vectors. Each element of the input vectors is compared element-wise, and the resulting logical vector will contain `TRUE` where the comparison is true and `FALSE` where it is not. This applies regardless of whether you are comparing vectors with each other or vectors with a single value. Here is an example of logical operations applied to vectors:

```
# Create a vector containing the values 1, 2, 3
vector_a <- c(1, 2, 3)

# Create a vector containing the values 4, 5, 6
vector_b <- c(4, 5, 6)

# Equality comparison of the two vectors
vector_a == vector_b
```

```
[1] FALSE FALSE FALSE
```

```
# Greater than comparison of the two vectors
vector_a > vector_b
```

```
[1] FALSE FALSE FALSE
```

```
# Less than or equal to comparison of the two vectors
vector_a <= vector_b
```

```
[1] TRUE TRUE TRUE
```

```
# Inequality comparison of the two vectors
vector_a != vector_b
```

```
[1] TRUE TRUE TRUE
```

You can combine logical operations using the AND (&) or the OR operator (|). Similar to arithmetic operators, these operators work element-wise and respect parenthetical grouping when used:

```
# AND operation on two logical vectors
(vector_a > 1) & (vector_b < 3)
```

```
[1] FALSE FALSE FALSE
```

```
# OR operation on two logical vectors
(vector_a == 1) | (vector_b > 2)
```

```
[1] TRUE TRUE TRUE
```

2.3.5 *Subsetting Vectors with Other Vectors*

We can use logical vectors to subset or filter elements from other vectors by providing a logical vector as the index. This will result in a new vector containing only the elements where the logical condition is `TRUE`. For example, let's say you have a vector of numbers and want to extract numbers less than 5. You can create a logical vector using a less-than comparison and use that as the original vector's index:

```
# Create a numerical vector
numerical_vector <- c(1, 4, 7, 3, 8, 2)

# Create a logical vector using the less-than comparison
less_than_5 <- numerical_vector < 5

# Subset the number_vector using the logical vector (1, 4, 3, 2)
numerical_vector[less_than_5]
```

```
[1] 1 4 3 2
```

Rather than creating intermediate vectors, you can perform these operations in a single statement. Here is the use of a single statement to subset the elements:

```
# Subset the number_vector using a less-than comparison directly (1, 4,
↪ 3, 2)
numerical_vector[numerical_vector < 5]
```

```
[1] 1 4 3 2
```

Like before, we can also use multiple logical conditions to subset a vector by combining them with the AND (&) or the OR operator (|). For instance, if we want numbers greater than 3 but less than 5:

```
# Subset the number_vector using multiple logical conditions
numerical_vector[(numerical_vector > 3) & (numerical_vector < 5)]
```

```
[1] 4
```

2.4 Functions

Functions are blocks of code that take inputs (called arguments) and return outputs. You can think of them as recipes in a cookbook. Just as a recipe provides step-by-step instructions on how to create a dish, a function provides a set of instructions to perform a specific task or calculation. Instead of writing the same code over and over again, you can define a function once and then call it whenever you need to perform that specific task. This helps reduce repetition and makes your code more concise and modular.

Functions have inputs, called arguments, which are the ingredients you provide to the recipe. These arguments are variables or values that the function will operate on or use in its calculations. For example, a function that calculates the average of a list of numbers would take the list as an argument.

Once the function has processed the inputs and performed the necessary calculations or actions, it can return an output. This output represents the result of the function's operation and can be used in other parts of your code. Going back to the recipe analogy, the output of a function is like the finished dish that you can serve or use in another recipe.

2.4.1 Calling Built-In Functions

R has numerous handy built-in functions that we can call by using their name followed by parentheses. We provide any necessary inputs (arguments) within these parentheses. We have already used functions earlier in this tutorial without noticing—for example, the `c()` function we used to create vectors. Some functions return a value, while others are used for their side effects, such as printing something to the console or modifying elements. Some common functions in R are `sum()`, `mean()`, `median()`, and `sd()` (standard deviation).

Let's examine how some of these functions work with a simple vector of numbers:

```
# Create a numeric vector
numerical_vector <- c(1, 2, 3, 4, 5)

# Calculate the sum of a numeric vector
sum(numerical_vector)
```

```
[1] 15
```

```
# Calculate the mean (average) of a numeric vector
mean(numerical_vector)
```

```
[1] 3
```

```
# Calculate the median of a numeric vector
median(numerical_vector)
```

```
[1] 3
```

```
# Calculate the standard deviation of a numeric vector
sd(numerical_vector)
```

```
[1] 1.581139
```

We can also use functions that take character vectors as their input, such as `paste()`, to concatenate strings together with a space or a specified separator:

```
# Create two character vectors
greeting <- "Hello"
name <- "John"

# Combine vectors together with a separator
paste(greeting, name, sep=", ")
```

```
[1] "Hello, John"
```

2.4.2 *Factors*

Factors are a data structure in R specifically designed to handle categorical data, which consist of a limited set of distinct values known as levels. They are particularly useful when dealing with non-numeric data that falls into specific categories, such as "Yes" and "No." In practical terms, converting character vectors to factors enables R to treat the data appropriately during statistical analysis and plotting. We can create factors by simply using the `factor()` function:

```
# Create a character vector with "Yes" and "No" responses
responses <- c("Yes", "No", "Yes", "No", "No", "Yes")

# Convert the character vector to a factor
response_factor <- factor(responses)

# Print the factor to see its levels
response_factor
```

```
[1] Yes No  Yes No  No  Yes
Levels: No Yes
```

As you can see, when we print a factor, we see different "levels" representing all the unique values present in the vector. By default, R will order the levels alphabetically unless otherwise specified. One reason you may want to change the order of levels is when your categorical data is ordered, such as "Low," "Medium," and "High." We can add an argument to the `factor()` function to specify an order for the levels by passing a vector with the levels in our desired order:

```
# Create a character vector with "Low", "Medium", and "High" responses
responses <- c("Medium", "Low", "High", "Medium", "Low")

# Create an ordered factor by specifying the order of levels
ordered_responses <- factor(responses, levels = c("Low", "Medium",
↪  "High"))

# Print the ordered factor to see its levels
ordered_responses
```

```
[1] Medium Low    High   Medium Low
Levels: Low Medium High
```

This adjusts the factor so that "Low" is considered the first level, "Medium" the second, and "High" the third, according to the order that we specified. We may also want to indicate that this factor is an ordered factor by including the `ordered =` TRUE argument, which tells R and anyone reading the code that the levels should be considered in a specific sequence. This can impact certain statistical analyses, such as ordinal regression, where the order of the factor levels can change the interpretation of the model.

One tricky issue you may run into is when categorical values are represented using numbers, such as for an identifier or a rating scale. You might think that R will automatically recognize these as categorical, but by default, it will treat any number as numeric data. If you want R to handle these numerical categories as categorical data, you'll need to convert them into factors, just as we did with character strings. For instance, let's imagine we are working on a dataset that has students divided into three groups labeled "1," "2," and "3." To ensure that these group labels are treated as categorical rather than numerical, we can convert them to factors as shown below:

```
# Create a vector of numerical group labels
group_labels <- c(1, 2, 3, 1, 3, 2, 1, 2, 3)

# Convert the numerical group labels into a factor
group_factors <- factor(group_labels, levels = c(1, 2, 3))

# Print the factor to verify that it has been converted correctly
group_factors
```

```
[1] 1 2 3 1 3 2 1 2 3
Levels: 1 2 3
```

2.4.3 Creating Custom Functions

The real power of functions comes from their ability to perform custom actions that may not be covered by built-in functions or to combine multiple actions into a single reusable function. In R, you can create custom functions using the `function()` keyword, followed by a set of instructions enclosed within a pair of curly braces ({ }). Similar to assigning values to variables, you can assign a function to a variable name.

Let's practice by creating a custom function called square_root_and_round that takes a numeric value as input, calculates its square root, and rounds the result to the nearest whole number. To do this, we define the function with the `function()` keyword, followed by the parameter (input) name within parentheses. Inside the function body, we calculate the square root of the input value using the `sqrt()` function and store the result in a variable. Next, we round the result using the `round()` function and store the rounded value in another variable. Finally, we use the `return()` function to specify the output of our custom function, which is the rounded value:

```
# Create a custom function called square_root_and_round
square_root_and_round <- function(x) {
  result <- sqrt(x)
  rounded_result <- round(result)
  return(rounded_result)
}

# Call the function we just created with "50" as the argument
square_root_and_round(50)
```

```
[1] 7
```

As you can see, when we call the custom function `square_root_and_round()`, it returns the rounded_result variable. R will actually automatically return the value of the last calculation, which in this case is the rounded result. This is a unique feature of R—you don't need to explicitly use the `return()` function to specify the output of your custom function. It simplifies the code by eliminating the need for an explicit return statement in most cases. However, there may be situations where you want to include the `return()` statement to explicitly indicate the output or to control the flow of the function. In general, though, in simple functions like the one we just created, it is not necessary to include `return()` unless you have a specific reason to do so. Thus, we can simplify our code as follows:

```
# Create a custom function called square_root_and_round without
↪  specifying an explicit return
square_root_and_round <- function(x) {
  result <- sqrt(x)
  round(result)
}
```

```
# Call the function we just created with "50" as the argument
square_root_and_round(50)
```

```
[1] 7
```

As you can see, we achieved the same result even without the `return()` statement.

2.4.4 Custom Functions with Multiple Arguments

Custom functions can take multiple arguments, which are particularly useful for performing more complex calculations and operations. When defining a custom function, you can specify multiple arguments by separating them with commas. Each argument inside the function can be used like a local variable, providing the function with the flexibility to work with different types of data and to carry out a variety of tasks.

Let's create an example to illustrate this concept. Consider a custom function called power_and_round that takes two numeric inputs: x and y. In this function, we raise the first input x to the power of the second input y and then round the result to the nearest whole number. By defining this function with two arguments, we can create a reusable code block that performs this specific calculation for any given pair of numeric inputs.

To call this custom function then, we provide the required inputs within the parentheses, separated by commas:

```
# Create a custom function called power_and_round
power_and_round <- function(x, y) {
  result <- x^y
  round(result)
}

# Call the power_and_round function with different values
power_and_round(2.1, 3.2)
```

```
[1] 11
```

```
power_and_round(5.1, 2.2)
```

```
[1] 36
```

```
power_and_round(10, 0.5)
```

```
[1] 3
```

Note, the order of the arguments matters. If you want to use a different order, you will need to specify the argument names when calling the function:

```
# Call the power_and_round function with input values of 2.1 and 3.2,
↪  specifying argument names
power_and_round(y = 3.2, x = 2.1)
```

```
[1] 11
```

```
# Call the power_and_round function with input values of 5.1 and 2.2,
↪  specifying argument names
power_and_round(y = 2.2, x = 5.1)
```

```
[1] 36
```

```
# Call the power_and_round function with input values of 10 and 0.5,
↪  specifying argument names
power_and_round(y= 0.5, x = 10)
```

```
[1] 3
```

In some cases, it can be useful to assign default values to the arguments in your custom functions. Using default values in your function arguments can make your code more flexible and user-friendly, as it allows users to easily customize the behavior of the function without always having to provide values for every single argument. To specify a default value for an argument, you can use the equal sign (=) operator within the function definition.

For example, let's modify our power_and_round function to include a default value for the y argument:

```
# Create a custom function called power_and_round with a default value
power_and_round <- function(x, y = 2) {
  result <- x^y
  round(result)
}
```

Now the function can be called with either one argument or two. When only "x" is specified, "y" will automatically be set to 2:

```
# Call the power_and_round function (y uses default value of 2)
power_and_round(12.5)
```

```
[1] 156
```

2.4.5 *Variable Scopes in Custom Functions*

Variable scope refers to the accessibility of variables in different parts of the code. In R, and most programming languages, variables defined within a function are local to that function. This means that the variables are only accessible within the function and do not affect or interfere with variables outside the function, even if they share the same name.

Let's consider an example:

```
# Create a global variable
x <- 10

# Define a function that uses a local variable with the same name
test_scope <- function() {
  x <- 5
  return(x)
}

# Print original variable value

print(x) # 10
```

```
[1] 10
```

```
# Print local variable inside function

test_scope()
```

```
[1] 5
```

You may have expected the value of "x" outside the function to change, but as demonstrated, the scope of a variable inside a function is limited to that function alone. The "x" inside `test_scope()` is not the same "x" that was defined outside the function. Each "x" variable exists in a separate environment; thus, when we modify "x" inside the function, it does not affect the global "x."

2.4.6 *Nested Functions and the Pipe Operator*

We can also nest functions. For instance, when we created our square root and rounding function above, we first calculated the square root and then applied rounding to the result. In the process, we created multiple intermediate variables, but instead of storing intermediate results in separate variables, we could have performed operations directly on the output of previous calculations. This reduces the memory overhead of creating and storing unnecessary variables, which can be particularly beneficial when working with large datasets or computationally intensive tasks.

Let's rewrite our `square_root_and_round()` function with nesting:

```
# Rewrite the square root and round function using nesting
square_root_and_round <- function(x) {
  round(sqrt(x))
}

# Call the function we just created with "15" as the argument
square_root_and_round(15)
```

```
[1] 4
```

While nesting works, it can be difficult to read when dealing with more complex calculations. In R, the pipe operator (|>) provides the ability to chain multiple functions together by passing the output of one function as the input of the next. To use the pipe operator, you simply place it between the two functions you want nested:

```
# Rewrite the square root and round function using the pipe operator
square_root_and_round <- function(x) {
  sqrt(x) |>
    round()
}

# Call the function we just created with "15" as the argument
square_root_and_round(15)
```

```
[1] 4
```

2.4.7 Alternative Pipe Operators and Installing Packages

When exploring code online, you may see an alternative pipe operator used in the R community: the "magrittr" pipe (%>%). This is because the base pipe operator was only released with R 4.1 in response to the "magrittr" package's popularity (Álvarez 2021). Throughout this book, we will use the base R pipe operator (|>) for consistency and simplicity, but feel free to explore and use the "magrittr" pipe operator (%>%) as you become more familiar with R. In most cases, the %>% pipe works similar to the base R pipe operator (|>). However, there are some instances where the behavior of the two pipes can differ slightly, due to some advanced features provided by the %>% pipe. To learn more about these, you can check the documentation of the "magrittr" package (Bache et al. 2022).

One of the benefits of the base R pipe operator is that it does not require any additional packages to be installed, making it universally accessible to any R user with the appropriate version of R. In comparison, the %>% pipe requires us to install the "magrittr" package. In R, installing packages is a simple process that can be done from the Comprehensive R Archive Network (CRAN) using the function `install.packages()`:

````
```{r}
# Install the magrittr package
install.packages('magrittr')
```
````

For the rest of the book, you will encounter various examples and exercises where you'll need to install packages to access different functions and enhance R's capabilities. You only need to do this the first time that you use a package on your computer, as once installed, you can just load it with the `library()` function in each new R session where you need it. Thus, now that you have installed the "magrittr" package, you can proceed to load it:

```
# Load the magrittr package
library('magrittr')
```

For the sake of demonstration, let's rewrite our `square_root_and_round()` function with the magrittr operator:

```
# Rewrite the square root and round function using the magrittr pipe
↪ operator
square_root_and_round <- function(x) {
  sqrt(x) %>%
    round()
}

# Call the function we just created with "15" as the argument
square_root_and_round(15)
```

```
[1] 4
```

2.5 Matrices

Matrices are two-dimensional data structures that organize data into rows and columns. They are similar to tables or spreadsheets. In a matrix, each element, or cell, holds a single value and can be accessed using its row and column indices. The indices start from 1 and increment by one for each subsequent row or column. This allows for easy retrieval and modification of specific data points within the matrix. One important characteristic of matrices is that they can only contain elements of the same data type. This means that all values in a matrix must be of the same type, such as integers, characters, or logical values.

To create a matrix in R, you can use the `matrix()` function. This function takes in the data elements as its input and allows you to specify the number of rows and columns in the matrix. Let's start by creating a simple matrix with two rows and three columns:

```
# Create a 2 x 3 matrix
my_matrix <- matrix(c(1, 2, 3, 4, 5, 6), nrow = 2, ncol = 3)

# Print the matrix to see its structure
my_matrix
```

```
     [,1] [,2] [,3]
[1,]    1    3    5
[2,]    2    4    6
```

As you can see, when we construct the matrix, R fills it column-wise by default. That is, it will start filling in the first column, then move on to the second, and so on. If instead, you want to fill the matrix by rows, you can add an additional argument `byrow = TRUE` to the `matrix()` function:

```
# Create a 2x3 matrix by filling values by rows
my_matrix <- matrix(c(1, 2, 3, 4, 5, 6), nrow = 2, ncol = 3, byrow =
↪  TRUE)

# Print the matrix to see its structure
my_matrix
```

```
     [,1] [,2] [,3]
[1,]    1    2    3
[2,]    4    5    6
```

2.5.1 Accessing Elements of a Matrix

To access matrix elements, use square brackets `[]` and provide the row and column indices separated by a comma. If you want to access an entire row or column, you can leave the corresponding index empty. This concept may be familiar to users who have worked with spreadsheets, as accessing cells in a spreadsheet is quite similar. Let's see how this works in practice:

```
# Access the element in the first row and second column
my_matrix[1, 2]
```

```
[1] 2
```

```
# Access the entire first row
my_matrix[1, ]
```

```
[1] 1 2 3
```

```
# Access the entire second column
my_matrix[, 2]
```

```
[1] 2 5
```

2.5.2 Matrix Arithmetic

As with vectors, arithmetic operations on matrices in R are performed element-wise. For instance, when performing the addition of two matrices, the value in row 1 and column 1 of the first matrix is added to the value in row 1 and column 1 of the second matrix, and the resulting sum is placed in the corresponding position of the resulting matrix. This element-wise operation is carried out for all elements in the matrices.

This concept of element-wise operations applies not only to addition but also to subtraction, multiplication, and division. Each element in the resulting matrix is computed by applying the corresponding arithmetic operation to the corresponding elements in the input matrices:

```
# Create two matrices

matrix_a <- matrix(c(1, 2, 3, 4, 5, 6), nrow = 3, byrow = TRUE)
matrix_b <- matrix(c(6, 5, 4, 3, 2, 1), nrow = 3, byrow = TRUE)

# Element-wise addition
matrix_a + matrix_b
```

```
     [,1] [,2]
[1,]    7    7
[2,]    7    7
[3,]    7    7
```

```
# Element-wise subtraction
matrix_a - matrix_b
```

```
     [,1] [,2]
[1,]   -5   -3
[2,]   -1    1
[3,]    3    5
```

```
# Element-wise multiplication
matrix_a * matrix_b
```

```
     [,1] [,2]
[1,]    6   10
[2,]   12   12
[3,]   10    6
```

```
#Element-wise division
matrix_a / matrix_b
```

```
          [,1]     [,2]
[1,] 0.1666667 0.400000
[2,] 0.7500000 1.333333
[3,] 2.5000000 6.000000
```

If you have a background in linear algebra, you may already be familiar with the concept of matrix multiplication. In R, matrix multiplication can be performed using the %*% operator. However, it is important to note that matrix multiplication is different from element-wise multiplication.

When performing matrix multiplication, the number of columns in the first matrix must match the number of rows in the second matrix. This ensures that the matrices are compatible for multiplication. The resulting matrix will have the same number of rows as the first matrix and the same number of columns as the second matrix. For instance, if we attempt

```
# Linear algebra style matrix multiplication
matrix_a %*% matrix_b
```

```
Error in matrix_a %*% matrix_b: non-conformable arguments
```

As you can see, R gives us a dimension mismatch error since "matrix_a" has three rows and two columns, while "matrix_b" has the same dimensions. To perform matrix multiplication, we need a matrix that complements the shape of "matrix_a," such as "matrix_c" with dimensions 2×3. Let's create "matrix_c" and then do the matrix multiplication:

```
# Create a 2x3 matrix
matrix_c <- matrix(c(1, 3, 5, 2, 4, 6), nrow = 2, byrow = TRUE)

# Perform matrix multiplication
matrix_a %*% matrix_c
```

```
     [,1] [,2] [,3]
[1,]    5   11   17
[2,]   11   25   39
[3,]   17   39   61
```

As you can see, this operation yields a 3 × 2 matrix, which is the product of "matrix_a" and "matrix_c."

2.6 Introducing Data Frames

Data frames are the main data types in R. Just like matrices, they can be thought of as similar to spreadsheet tables, where each column represents a different variable or attribute, and each row represents an observation or data point. Unlike matrices, however, which require all elements to be of the same data type, data frames can contain columns of different types. This flexibility allows you to store and analyze heterogeneous data, such as numeric values, character strings, factors, and logical values, within a single data structure.

In addition to storing data, data frames also retain important metadata, such as variable names and column labels. This makes it easier to interpret and analyze the data, as well as perform operations like subsetting, merging, and aggregating data based on specific variables or conditions.

To create a data frame in R, you can use the `data.frame()` function. This function takes the variables as arguments, which should be provided in the form of vectors. Each vector will represent a column in the data frame, and the elements within each vector will be the values for that column. For example, let's say you have three variables: "Name," "Age," and "Salary":

```
# Create vectors for Name, Age, and Salary
Name <- c("Alice", "Bob", "Charlie")
Age <- c(25, 32, 28)
Salary <- c(50000, 48000, 58000)

# Create a data frame using the vectors
employee_data <- data.frame(Name, Age, Salary)

# Print the data frame
employee_data
```

```
     Name Age Salary
1   Alice  25  50000
2     Bob  32  48000
3 Charlie  28  58000
```

As you can see, when we printed the data frame, each variable becomes a separate column, and each observation becomes a row. We also got the name of the variables displayed as the headers of the columns, which provides a clear overview of what each column represents. This feature of data frames makes it much easier to manage and work with your data, especially when dealing with complex datasets that may contain hundreds or thousands of observations and variables.

2.6.1 Accessing Elements of a Data Frame

You can access specific rows or elements in a data frame using the square brackets [] notation you saw with matrices. For instance, to select the second row in "my_dataframe," you would use the following syntax:

```
# Access the second row of the data frame
employee_data[2, ]
```

```
  Name Age Salary
2  Bob  32  48000
```

This will output the entire second row of the data frame, which in our example contains Bob's data.

On the other hand, to subset by both rows and columns, you can specify the row index(es) and the column name(s). For instance, if you only wanted Bob's salary, you would use the following:

```
# Access Bob's Salary
employee_data[2, 3]
```

```
[1] 48000
```

Of course, remembering the name of the column might not always be practical, especially in data frames with many variables. Luckily, R allows you to access a column in three additional ways. First, you could use the name of the column in the brackets ([]) as a string:

```
# Access Age column by name
employee_data[, "Age"]
```

```
[1] 25 32 28
```

Second, you could also use the ($) operator, which allows you to extract a specific variable without quotes:

```
# Access Age column using the ($) operator
employee_data$Age
```

```
[1] 25 32 28
```

Third, you can also use double brackets (`[[]]`) to access a column by its index or name:

```
# Access Age column by index using double bracket notation
employee_data[[2]]
```

```
[1] 25 32 28
```

```
# Access Age column by name using double bracket notation
employee_data[["Age"]]
```

```
[1] 25 32 28
```

2.6.2 Subsetting Data Frames

We can use logical conditions to subset data frames in R, similar to how we use them with vectors. This allows us to focus our analysis on specific subsets of the data based on conditional criteria. For example, if we want to extract rows where the "Age" is greater than 30:

```
# Subset Employee Data to only see cases where Age is greater than 30
subset_data <- employee_data[employee_data$Age>30, ]
```

This will give us a smaller data frame with only the rows where the "Age" column has values greater than 30. What has occurred is that R evaluated the condition `employee_data$Age > 30` and returned a logical vector of the same length as the number of rows in our data frame. It then used this vector to choose which rows to include in the new data frame.

Just as before, we can use the AND (&) and OR (|) operators to combine multiple conditions when subsetting our data frame. For instance, if we need to find employees who are older than 30 and have a "Salary" greater than or equal to 60,000:

```
# Subset for Age > 30 and Salary >= 60000
higher_income_older_employees <- employee_data[(employee_data$Age > 30) &
  (employee_data$Salary >= 60000), ]
```

We can also subset for both rows and columns in a single operation by specifying row and column indices or names within the brackets. For example:

```
#Get the Names and Salaries of employees aged over 30
older_employees <- employee_data[employee_data$Age > 30, c("Name",
↪  "Salary")]
```

In this command, we tell R to select only the columns labeled "Name" and "Salary" for all rows where the "Age" is over 30, resulting in a table that focuses on the desired subset of staff information.

By applying these logical subsets, you can conduct more targeted data analyses. This is particularly useful when working with large datasets, since subsetting can significantly reduce the amount of data you need to process.

2.6.3 Adding and Removing Columns and Rows in Data Frames

Often, you will need to modify your data frame by adding new data or removing unnecessary information. To add a new column to a data frame, you can simply use the `$` operator and provide a vector of the same length as the number of rows in the data frame. For instance, let's say we want a new column in our data frame that consists of the years of experience each employee has:

```
# Add a new column to the dataframe showing employee experience
employee_data$Experience <- c(2, 5, 8)

# Print employee_data
employee_data
```

```
     Name Age Salary Experience
1   Alice  25  50000          2
2     Bob  32  48000          5
3 Charlie  28  58000          8
```

Alternatively, you can use the `cbind()` function to add an additional column. This function takes the name of the original data frame and the vector you want to add as its arguments:

```
# Add a new column to the dataframe using cbind()
employee_data <- cbind(employee_data, Department = c("HR", "IT",
↪  "Finance"))

# Print employee_data
employee_data
```

```
     Name Age Salary Experience Department
1   Alice  25  50000          2         HR
2     Bob  32  48000          5         IT
3 Charlie  28  58000          8    Finance
```

Just as we added columns, we can also add rows to a data frame with the rbind() function. Assuming you have a new employee data you'd like to add, you could do the following:

```
# Create new sample employee data
new_employee <- data.frame(Name="Dan",
                           Age=28,
                           Salary=48000,
                           Experience=1,
                           Department="Marketing")

#Add a new row with rbind()
employee_data <- rbind(employee_data, new_employee)

# Print updated employee_data
employee_data
```

```
     Name Age Salary Experience Department
1   Alice  25  50000          2         HR
2     Bob  32  48000          5         IT
3 Charlie  28  58000          8    Finance
4     Dan  28  48000          1  Marketing
```

This appends a new row containing Dan's information to "my_dataframe." Keep in mind, the names and number of columns in the data frame you're adding should match the existing data frame for `rbind()` to work correctly.

2.7 The Concept of Lists

Earlier, we learned that R has two main types of vectors: atomic and lists. While atomic vectors can only contain elements of the same type, lists provide a way to store a collection of objects of different types, lengths, and dimensions within a single variable. In other words, you can combine vectors, matrices, data frames, or even other lists, into a single list. Each object within the list is referred to as an element, and elements can be accessed using the double square bracket notation (`[[]]`).

To create a list in R, you can use the `list()` function. The arguments passed to this function are the elements of the list and you can name each element of the list if you wish by assigning a value to a named argument. For example, let's say you have a list called "my_list" that contains three elements: a numeric vector, a character vector, and a data frame. You can create the list as follows:

```
# Create sample numerical vector
my_numbers <- c(52, 32, 43)

# Create sample character vector
```

```
my_characters <- c("z", "j", "k")

# Create a sample data frame
my_dataframe <- data.frame(x = 23:25, y = c("L", "X", "R"))

# Create a list of the two vectors and data frame
my_list <- list(numbers=my_numbers,
                characters=my_characters,
                my_data=my_dataframe)

# Print my_list
my_list
```

```
$numbers
[1] 52 32 43

$characters
[1] "z" "j" "k"

$my_data
   x y
1 23 L
2 24 X
3 25 R
```

In this example, the "my_list" contains three elements: "numbers," "letters," and "data." Each element has its own name, which allows you to easily access specific elements within the list.

To access elements within a list, just as with data frames, you can use the double square bracket notation (`[[]]`) followed by the name of the element. For example, to access the "numbers" element in "my_list," you would use `my_list[["numbers"]]`. Similarly, you can access other elements by replacing "numbers" with the name of the desired element:

```
# Access the 'characters' element
my_characters <- my_list[["characters"]]

# Access the 'my_data' element
my_data <- my_list[["my_data"]]

# Print the 'my_data' data frame
my_data
```

```
   x y
1 23 L
2 24 X
3 25 R
```

Keep in mind that using the single square brackets `[]` notation will result in a sub-list with the selected elements, rather than extracting the individual elements themselves. This contrasts with the double brackets ([[]]), which don't return the list structure but rather the contents of the selected element directly. For instance:

```
# Return a sub-list containing the 'my_data' element
sub_list <- my_list["my_data"]

# Print the sublist to show its structure
sub_list
```

```
$my_data
   x y
1 23 L
2 24 X
3 25 R
```

2.7.1 The Relationship Between Data Frames and Lists

Lists and data frames in R have similarities in their structure. In fact, data frames are a special type of list with additional constraints. In a list, you can store elements of different types, lengths, and dimensions, making it a flexible data structure. Each element in a list can be accessed using indexing and can be of any length or type. On the other hand, data frames require all columns to have the same length and maintain a tabular structure.

For instance, let's say we are interested in converting our employee data frame to a list. You can do so by using the `as.list()` function:

```
# Convert the dataframe to a list
employee_list <- as.list(employee_data)

# Print the list to see its structure
employee_list
```

```
$Name
[1] "Alice"   "Bob"     "Charlie" "Dan"

$Age
[1] 25 32 28 28

$Salary
[1] 50000 48000 58000 48000

$Experience
[1] 2 5 8 1

$Department
[1] "HR"        "IT"        "Finance"   "Marketing"
```

After conversion, "employee_list" contains each column of "employee_data" as individual list elements. Now, suppose we want to convert back to a data frame—noting that all list elements are of the same length—we can use the `as.data.frame()` function:

```
# Convert the list back to a dataframe
employee_data <- as.data.frame(employee_list)

# Print the data frame to see its structure
employee_data
```

```
     Name Age Salary Experience Department
1   Alice  25  50000          2         HR
2     Bob  32  48000          5         IT
3 Charlie  28  58000          8    Finance
4     Dan  28  48000          1  Marketing
```

This round trip from a data frame to a list and back again illustrates how these two types of objects can be interchanged, given that the list has elements of the same length, addressing the requirements of a data frame's structure.

2.8 Reading in Datasets

You will rarely create datasets by hand. Instead, you will import data from external sources such as CSV files, Excel files, or databases. In this book, we will import our data as CSV (comma-separated value) files, which are arguably the most common data file format.

CSV files are plain text files that separate each value with a comma. The first row will contain the labels for the columns with each additional row separated by a new line and representing a new observation. Here is how a simple dataset of name, age, and salary would look:

```
Name, Age, Salary
John, 34, 54535
Sally, 53, 63234
Joe, 55, "70,000"
```

You may be wondering why Joe's salary is in quotes. This is because there is a comma in the value itself. The quotation marks indicate that the actual value for that column contains a comma and that the comma does not represent a new column.

2.8.1 Loading a CSV File

For the remainder of this chapter, we will work with a dataset compiled by Randi H. Griffen, which examines over 120 years of Olympic athletes from Athens 1896 to Rio 2016. This dataset allows us to explore various aspects of the Olympics, including health, popularity, nationalism, and gender dynamics over time (Griffin 2018).

To read a CSV file, we call the `read.csv()` function with an argument specifying the file's path. In the example below, along with the rest of the book, I assume you have downloaded the data folder from the online repository and placed it in your working directory. You can call `getwd()` to find out your current working directory and change it using the `setwd()` function, which takes the path to a new directory as its argument, if needed:

```
# Use getwd() to get your working directory
# Use setwd() if necessary to change your working directory

# Read the CSV file
olympics <- read.csv("data/ch1_olympics.csv")
```

You can see that R has automatically turned our dataset into a data frame by using the class() function:

```
# Check the class of the imported data
class(olympics)
```

```
[1] "data.frame"
```

Now that we have loaded our data, we can print its contents by calling it:

````
```{r}
olympics
```
````

```
Rows: 206,853
Columns: 15
$ id     <int> 33539, 33539, 33539, 33539, 33541, 33542, 33544, 33544, 33545, ~
$ name   <chr> "Bndicte Evrard", "Bndicte Evrard", "Bndicte Evrard", "Bndicte ~
$ sex    <chr> "F", "F", "F", "F", "M", "M", "F", "F", "F", "F", "F", "M", "M"~
$ age    <int> 17, 17, 17, 17, 21, 24, 19, 24, 21, 25, 29, 30, 30, 30, 24, 24,~
$ height <int> 176, 176, 176, 176, 185, 180, 176, 176, 173, 173, 173, 180, 180~
$ weight <dbl> 38, 38, 38, 38, 77, 76, 62, 62, 65, 65, 65, 71, 71, 71, 61, 61,~
$ team   <chr> "Belgium", "Belgium", "Belgium", "Belgium", "Germany", "Netherl~
$ noc    <chr> "BEL", "BEL", "BEL", "BEL", "GER", "NED", "FRA", "FRA", "FRA", ~
$ games  <chr> "1992 Summer", "1992 Summer", "1992 Summer", "1992 Summer", "20~
$ year   <int> 1992, 1992, 1992, 1992, 2004, 1964, 1984, 1988, 1988, 1992, 199~
$ season <chr> "Summer", "Summer", "Summer", "Summer", "Summer", "Summer", "Su~
$ city   <chr> "Barcelona", "Barcelona", "Barcelona", "Barcelona", "Athina", "~
$ sport  <chr> "Gymnastics", "Gymnastics", "Gymnastics", "Gymnastics", "Canoei~
$ event  <chr> "Gymnastics Women's Floor Exercise", "Gymnastics Women's Horse ~
$ medal  <chr> "None", "None", "None", "None", "None", "None", "None", "None",~
```

Note: Due to space restraints, this book uses an alternative printing style for data frames (as shown above using `dplyr::glimpse()`, a function from the `dplyr` package, which we will explore in the next chapter). For now, do not worry if your output does not look exactly the same as standard R printing.

2.8.2 Analyzing the Structure of a Dataset

Now that we have loaded our data, we can take a look at it. With large datasets, it is helpful to get a quick glance at the data. R has several ways to do this, but the most straightforward is the head() function, which displays the first six rows of the dataset:

```r
# View the first 6 rows of the dataset
head(olympics)
```

```
     id           name sex age height weight         team noc       games year
1 33539 Bndicte Evrard   F  17    176     38      Belgium BEL 1992 Summer 1992
2 33539 Bndicte Evrard   F  17    176     38      Belgium BEL 1992 Summer 1992
3 33539 Bndicte Evrard   F  17    176     38      Belgium BEL 1992 Summer 1992
4 33539 Bndicte Evrard   F  17    176     38      Belgium BEL 1992 Summer 1992
5 33541     Jens Ewald   M  21    185     77      Germany GER 2004 Summer 2004
6 33542 Jacques Ewalds   M  24    180     76 Netherlands NED 1964 Summer 1964
  season      city      sport                               event medal
1 Summer Barcelona Gymnastics     Gymnastics Women's Floor Exercise  None
2 Summer Barcelona Gymnastics        Gymnastics Women's Horse Vault  None
3 Summer Barcelona Gymnastics        Gymnastics Women's Uneven Bars  None
4 Summer Barcelona Gymnastics       Gymnastics Women's Balance Beam  None
5 Summer    Athina   Canoeing Canoeing Men's Kayak Singles, Slalom  None
6 Summer     Tokyo Volleyball           Volleyball Men's Volleyball  None
```

Similarly, we can use tail() to get the last six rows:

```r
# View the last 6 rows of the dataset
tail(olympics)
```

```
           id                          name sex age height weight          team
206848 118191               Leonds Tambijevs   M  35    176     85        Latvia
206849 129331 Douglas Daniel "Doug" Weight   M  35    180     91 United States
206850  16836       Giuseppe "Joe" Busillo   M  35    186     96         Italy
206851   4527    Vladimir Yuryevich Antipin   M  35    187     92    Kazakhstan
206852 107285      Mathieu David Schneider   M  36    180     85 United States
206853 104543     Joseph Steven "Joe" Sakic   M  36    180     88        Canada
       noc       games year season   city      sport
206848 LAT 2006 Winter 2006 Winter Torino Ice Hockey
206849 USA 2006 Winter 2006 Winter Torino Ice Hockey
206850 ITA 2006 Winter 2006 Winter Torino Ice Hockey
206851 KAZ 2006 Winter 2006 Winter Torino Ice Hockey
206852 USA 2006 Winter 2006 Winter Torino Ice Hockey
206853 CAN 2006 Winter 2006 Winter Torino Ice Hockey
```

```
                                 event medal
206848 Ice Hockey Men's Ice Hockey  None
206849 Ice Hockey Men's Ice Hockey  None
206850 Ice Hockey Men's Ice Hockey  None
206851 Ice Hockey Men's Ice Hockey  None
206852 Ice Hockey Men's Ice Hockey  None
206853 Ice Hockey Men's Ice Hockey  None
```

You can also retrieve basic information about the dataset, such as its dimensions or the names of its columns, using the `dim()` and `names()` functions:

```
# R Get the dimensions of the dataset (rows and columns)
dim(olympics)
```

```
[1] 206853     15
```

```
# Get the names of the columns
names(olympics)
```

```
 [1] "id"     "name"   "sex"    "age"    "height" "weight" "team"   "noc"
 [9] "games"  "year"   "season" "city"   "sport"  "event"  "medal"
```

As we see, we have 206,853 athletes with 15 columns/variables for each:

id: A unique identification number assigned to each athlete
name: The name of the athlete
sex: The binary sex of the athlete (M for male, F for female)
age: The age of the athlete in years
height: The height of the athlete in centimeters
weight: The weight of the athlete in kilograms
team: The name of the team the athlete belongs to
noc: The National Olympic Committee's three-letter code representing the athlete's country.
games: The year and season of the Olympic Games the athlete competed in
year: The specific year of the Olympic Games the athlete competed in
season: The season of the Olympic Games the athlete competed in (Summer or Winter)
city: The host city of the Olympic Games the athlete competed in
sport: The sport in which the athlete competed
event: The specific event within the sport in which the athlete competed
medal: The type of medal won by the athlete (Gold, Silver, Bronze) or NA if no medal was won

With this information, we can start conducting different types of analyses to understand how different factors influence athlete performance or participation rates over time.

2.9 Calculating Summary Statistics

When first looking at a dataset, it is helpful to calculate summary statistics. You are probably used to seeing summary statistics, such as averages, minimums, and maximums, through media headlines. Since R is geared toward statistical analysis, it has easy built-in functions that handle these calculations. Let us get the mean, minimum, and maximum heights for each player:

```
# Get Mean Height
mean(olympics$height)
```

```
[1] 175.3591
```

```
# Get Minimum Height
min(olympics$height)
```

```
[1] 127
```

```
# Get Maximum Height
max(olympics$height)
```

```
[1] 226
```

We see that the mean height is 175.3591 cm, the minimum height is 127 cm, and the maximum height is 226 cm. Such statistics provide us with a quick understanding of the range and central tendency within the heights of Olympic athletes, allowing further analysis on how an athlete's height might influence their performance in different sports. However, summary statistics on their own do not tell us everything. For deeper insights, we might want to consider the height in relation to other variables such as the sport they participate in or the country they represent.

As before, we can subset our data to compare different groups. Let us compare the United States with Canada. I will be using the pipe operator (|>) in this example to make the code more readable. Remember that the pipe operator, a base R feature since version 4.1, simply allows us to send what is to the left of the pipe as the first argument to the function on the right. This will be very useful in the next chapter, so getting some practice right now will help. Make sure to review earlier in the chapter if you have not understood it fully:

```
# Filter the 'height' column for the United States and Canada
filtered_US <- olympics$height[olympics$team=="United States"]
filtered_Canada <- olympics$height[olympics$team=="Canada"]

# Calculate the mean, minimum, and maximum height for the United States
filtered_US |> mean()
```

```
[1] 176.9068
```

```
filtered_US |> min()
```

```
[1] 139
```

```
filtered_US |> max()
```

```
[1] 223
```

```
# Calculate the mean, minimum, and maximum height for Canada
filtered_Canada |> mean()
```

```
[1] 174.8844
```

```
filtered_Canada |> min()
```

```
[1] 132
```

```
filtered_Canada |> max()
```

```
[1] 216
```

We see that the mean height for athletes from the United States is slightly higher than that for Canadian athletes. However, the range of heights is quite similar between the two countries, which indicates that height may not be a significant differentiator between athletes from these regions.

Calculating summary statistics for individual columns can be tedious. Luckily, R provides a way to easily get summary statistics for all variables in a dataset through the `summary()` function, which takes the name of a data frame as its argument:

```
# Get summary of all columns both categorical and numerical
summary(olympics)
```

```
       id              name               sex                 age       
 Min.   :     1   Length:206853      Length:206853      Min.   :11.00  
 1st Qu.: 35141   Class :character   Class :character   1st Qu.:21.00  
 Median : 68595   Mode  :character   Mode  :character   Median :24.00  
 Mean   : 68583                                         Mean   :25.06  
 3rd Qu.:102283                                         3rd Qu.:28.00  
 Max.   :135571                                         Max.   :71.00  
                                                        NA's   :688    
     height          weight            team               noc           
 Min.   :127.0   Min.   : 25.00   Length:206853      Length:206853     
 1st Qu.:168.0   1st Qu.: 60.00   Class :character   Class :character  
 Median :175.0   Median : 70.00   Mode  :character   Mode  :character  
```

```
 Mean   :175.4   Mean   : 70.68
 3rd Qu.:183.0   3rd Qu.: 79.00
 Max.   :226.0   Max.   :214.00

    games                year         season              city
 Length:206853      Min.   :1896   Length:206853      Length:206853
 Class :character   1st Qu.:1976   Class :character   Class :character
 Mode  :character   Median :1992   Mode  :character   Mode  :character
                    Mean   :1990
                    3rd Qu.:2006
                    Max.   :2016

    sport               event               medal
 Length:206853      Length:206853      Length:206853
 Class :character   Class :character   Class :character
 Mode  :character   Mode  :character   Mode  :character
```

2.9.1 *Working with Summary Statistics for Categorical Columns*

In the `summary()` call above, we did not get very useful information for our categorical columns. For instance, you may think that running `maximum()` and `minimum()` on the medal column would provide the most and least common medals. However, functions like `mean()`, `min()`, and `max()` only work with numerical variables. To get counts for categorical variables, you need to use the `table()` function, which creates a contingency table. Let us look at how many Olympic athletes received medals:

```
# Get contingency table of how many medals there are
table(olympics$medal)
```

```
Bronze   Gold   None Silver
 10154  10168 176657   9874
```

Just as with descriptive statistics, we can subset our data frame, which allows us to ask more directed questions, such as how many medals each country has won. Let us compare medals won by China and Russia:

```
# Filter the Olympics dataframe to include only rows where the team is
↪  either "Russia" or "China"
olympics_russia_china <- olympics[olympics$team=="Russia" |
↪  olympics$team=="China", ]
```

```
# Create a contingency table showing the counts of medals for each team
table(olympics_russia_china$team, olympics_russia_china$medal)
```

```
         Bronze Gold None Silver
  China     266  308 3815    323
  Russia    392  356 3528    343
```

We see that Russia has a higher count of gold medals, which could be interesting for further investigation. We might also analyze other factors such as the number of athletes per country, or the sports in which these medals were won, to provide additional context.

We may also be interested in looking at proportions rather than raw counts. We can do this by passing our contingency table to the `prop.table()` function:

```
# Get proportions rather than raw counts of our contingency table
table(olympics_russia_china$team, olympics_russia_china$medal) |>
  prop.table()
```

```
             Bronze       Gold       None     Silver
  China  0.02850713 0.03300825 0.40885221 0.03461580
  Russia 0.04201050 0.03815240 0.37809452 0.03675919
```

2.10 Control Structures

Conditional statements and loops are fundamental programming concepts that allow us to automate tasks in R. Conditional statements enable our code to make decisions and perform different actions based on whether a condition is true or false. Loops, on the other hand, allow us to repeat a block of code for a set number of times or until a specific condition is met.

Because R is optimized for vectorized operations, where operations are performed on entire vectors rather than individual elements, loops are less commonly used in R compared to other programming languages. Vectorization allows for more efficient and faster computations when working with large datasets or performing complex calculations. However, there are still cases where control structures, such as if, else, for, and while loops, can be useful in R.

We will see how these operations function through the Olympics dataset. Understanding these will be particularly beneficial if you decide to branch into other programming languages later on, as these concepts are universal.

2.10.1 Using "If" Statements

An if statement is a control structure that allows a block of code to be executed when a specified condition is met. It provides a way to make decisions in your code based on certain criteria. For example, if you want to print a message when the mean height of athletes in the Olympics is greater than 175 cm, you can use an if statement. In this case, you would use the `message()` function to display the message you want to print as a result. Let's take a look at how this works:

```
# Create a variable that holds the mean height
mean_height <- mean(olympics$height)

# Check if the mean height is greater than 175cm and print a message if
↪   it is
if (mean_height > 175) {
  message("The mean height of athletes is greater than 175 cm.")
  }
```

Sometimes, you may also want to include code that runs when a condition is not met. In these cases, you can use the "else" clause in your if statement. For example, let's say we want to check if the average height is greater than 180 cm:

```
# Check if the mean height is greater than 180cm and print a message if
↪   it is
if (mean_height > 180) {
  message("The average height of athletes is greater than 180 cm.")
} else {
  message("The average height of athletes is not greater than 180 cm.")
}
```

In this code, if the condition (average_height > 180) evaluates to `TRUE`, the first block of code will execute and the message "The average height of athletes is greater than 180 cm." will be printed. Otherwise, if the condition is `FALSE`, the second block of code will execute and the message "The average height of athletes is not greater than 180 cm." will be printed.

Certainly! Here's an expanded version of the paragraph, along with the code comments:

A while loop allows you to repeat a block of code as long as a specified condition is true. For example, let's say we want to print the first ten unique sports. To do this, we need to keep track of a count variable and use the `unique()` function to get a vector of unique sports from our Olympics dataset. Then, we can use a while loop to repeat the code block as long as the count is less than ten. Inside the loop, we will print the current sport and increment the count. Once the count reaches ten, the loop will terminate and we will have printed the first ten unique sports:

```
# Initialize the count variable
i <- 1

# Get a vector of unique sports from the Olympics dataset
unique_sports <- unique(olympics$sport)

# Use a while loop to print the first 10 unique sports
while (i <= 10) {
  message(unique_sports[i])
  i <- i + 1
}
```

2.10.2 *For Loops*

In many cases, you will need to iterate over a set of items, such as a vector or a sequence of numbers, as you did with the while loop. A for loop in R allows you to do this more concisely and efficiently. In a for loop, you specify a variable and a sequence to iterate over. The loop will execute the code block for each item in the sequence, with the variable taking on the value of each item in turn. This allows you to perform repetitive tasks or calculations on each element of the sequence without having to manually increment the variable. The for loop automatically handles the iteration for you, making your code more readable and eliminating the need for extra code to track the index or iteration count:

```
# Get a vector of unique sports from the Olympics dataset
unique_sports <- unique(olympics$sport)

# Use a for loop to print the first 10 unique sports
for (i in 1:10) {
  message(unique_sports[i])
}
```

As you can see, we receive the same output, but the for loop is simpler and does not require us to manually increase the counter.

2.10.3 *The Apply Family of Functions*

As mentioned earlier, R is optimized for vectorized operations. Rather than for loops, you will find R code often utilizes the "apply" set of functions. Some of the primary apply functions in R include the following:

apply(): Applies a function on array or matrix margins
lapply(): Applies a function on each element of a list or vector and returns a list
sapply(): Similar to lapply(), but returns a vector, matrix, or array
tapply(): Applies a function on subsets of a vector, grouped by factors
mapply(): Applies a function on multiple lists or vectors, element-wise

These apply functions can greatly simplify and speed up code when working with large datasets or complex calculations. We will focus on the tapply() function in this chapter, as it is particularly useful in aggregating and summarizing data by groups. The rest of the apply family functions similarly and can be explored in other resources available online.

The tapply() function specifically allows you to apply a function to subsets of a vector or array, with the subsets defined by a grouping factor. This function takes three arguments: the input vector or array, the grouping factor, and the function you want to apply. For example, let's say we want to calculate the average height of athletes within each sport; we can use the tapply() function. First, we select the height variable from our dataset as the input vector. Then, we specify the grouping factor as the sport variable. Finally, we pass the mean() function as the function to apply:

```
# Calculate the mean height for each sport in the Olympics dataset
tapply(olympics$height, olympics$sport, mean)
```

```
            Alpine Skiing                   Archery          Art Competitions
                 173.5177                  173.2505                  172.8214
                Athletics                 Badminton                  Baseball
                 176.2596                  174.2203                  182.5993
               Basketball          Beach Volleyball                  Biathlon
                 191.2090                  186.2026                  174.0520
                Bobsleigh                    Boxing                  Canoeing
                 181.6043                  172.7885                  178.5521
     Cross Country Skiing                   Curling                   Cycling
                 173.2613                  174.1683                  176.1624
                   Diving             Equestrianism                   Fencing
                 166.6622                  174.3971                  177.2983
           Figure Skating                  Football          Freestyle Skiing
                 168.1488                  175.3875                  170.9409
                     Golf                Gymnastics                  Handball
                 174.0463                  162.8756                  183.4111
                   Hockey                Ice Hockey                      Judo
                 173.3879                  178.9231                  174.1572
                 Lacrosse                      Luge         Modern Pentathlon
                 174.0000                  176.6550                  178.1015
              Motorboating           Nordic Combined      Rhythmic Gymnastics
                 181.0000                  176.5197                  167.7951
                   Rowing                     Rugby              Rugby Sevens
                 184.2626                  173.6667                  175.3636
                  Sailing                  Shooting Short Track Speed Skating
                 178.2638                  173.5268                  170.1082
                 Skeleton               Ski Jumping               Snowboarding
```

```
              176.2241                  176.4898              173.0433
              Softball             Speed Skating              Swimming
              169.4286                  174.0969              178.5901
 Synchronized Swimming              Table Tennis             Taekwondo
              168.4900                  171.2538              176.7500
                Tennis               Trampolining            Triathlon
              178.9158                  166.5638              173.6407
            Tug-Of-War                Volleyball           Water Polo
              182.7273                  186.9789              185.0283
         Weightlifting                 Wrestling
              167.8227                  172.3894
```

By using the apply functions, you can avoid writing explicit for loops and instead apply a function to subsets of your data in a single, concise line of code. The benefits of vectorized operations are particularly evident in data science tasks, where you often need to perform calculations on entire datasets. Whether you're calculating summary statistics, manipulating data, or performing complex calculations, using the apply functions can greatly simplify your code and make it more efficient.

While vectorized operations may take some time to get used to, they are a powerful feature of R that can significantly enhance your data analysis and statistical modeling. They are one of the key reasons why R is a popular language for data science and why it's widely used in the research and industry communities.

2.11 Basic Plotting in R

Visualizations are very helpful for cultural analysis because they enable us to see patterns and relationships in the data that may not be obvious from raw numbers alone. In Chap. 3, we will look at the "ggplot2" package and its theoretical underpinnings. If you look online, you will find that "ggplot2" has become the de facto plotting library in R. This has advantages and disadvantages. "ggplot2" requires a strong foundation in an abstract framework known as the "Grammar of Graphics," which can be challenging to learn while also learning syntax. The base R library for graphics, on the other hand, is relatively straightforward while still powerful.

The default values for base R plots are sufficient for basic visualizations, but you can also modify aspects such as labels, colors, and titles with additional arguments. In this section, we will stick with creating simple plots, leaving more advanced plot customizations for the "ggplot2" chapter. Specifically, we will cover how to create histograms, bar plots, and scatter plots using base R graphics.

2.11.1 Creating Histograms

Histograms visualize the distribution of data and are useful in understanding shape, skewness, and potential outliers in a dataset. In R, you can create a histogram using the `hist()` function. The primary argument for this function is the input vector containing the data. For instance, if we wanted to draw a histogram of athlete heights in our Olympics dataset, we could do the following:

```
# Create a histogram showing the height distribution of olympic athletes
hist(olympics$height)
```

As shown in Fig. 2.1, it seems that the distribution of athlete weights follows a bell curve, meaning most athletes cluster around a median weight with fewer athletes representing extremely high or low weight values. This type of distribution is common in biological data and is known as a normal distribution. However, is this actually the case?

By default, R will display the histogram with a reasonable number of bins based on the input. However, you can customize the number of bins or the breaks between bins using the `breaks` argument. For example, to create a histogram with 40 bins for athlete weights, we could do the following:

```
# Create a histogram showing the weight distribution of olympic athletes
↪  with 40 bins
hist(olympics$weight, breaks=40)
```

Figure 2.2 shows that this updated histogram has a positive or right skew. This indicates that the average athlete's weight is slightly less than the median, suggesting that while most athletes are of moderate weight, there's a significant number of athletes whose weight is above the average, creating the tail on the right side. This

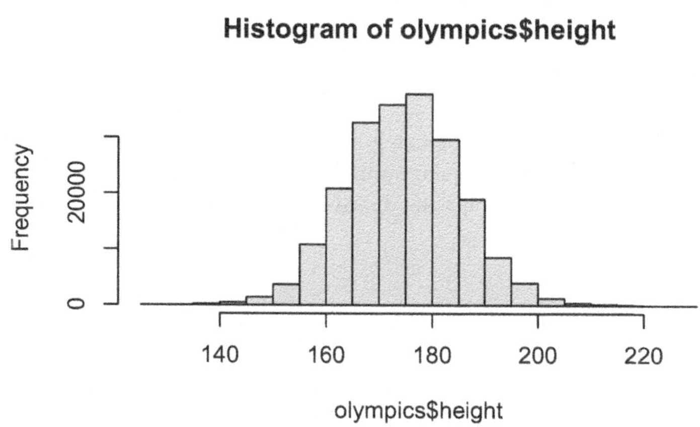

Fig. 2.1 Distribution of heights among Olympic athletes

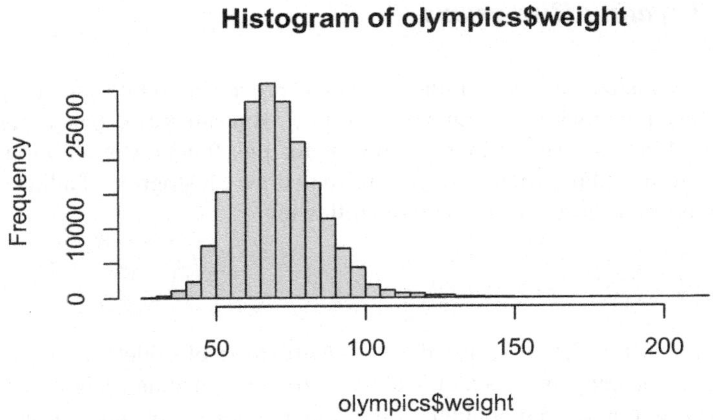

Fig. 2.2 Distribution of weights among Olympic athletes, using 40 bins

could be due to the inclusion of certain sports where heavier weights are more common, such as weightlifting. It's this kind of insight that can guide more specific research questions or inform data preprocessing decisions in future analyses.

2.11.2 *Making Bar Plots*

Bar plots visualize the counts of categorical data and are arguably the most common type of plot used to represent qualitative information. In R, you can create a bar plot using the `barplot()` function, which takes a contingency table as its input. For instance, to create a bar plot showing the number of medals in our Olympics dataset, we can do the following:

```
# Create a bar plot showing the medal counts
barplot(table(olympics$medal))
```

Figure 2.3 displays counts of categorical data; do not confuse bar plots with histograms. While they may appear similar, bar plots display counts of categorical data, whereas histograms display the distribution of continuous data.

2.11.3 *Scatter Plots*

Scatter plots visualize the relationship between two continuous variables and can identify trends and correlations in data. In R, you create a scatter plot using the `plot()` function. The primary arguments for this function are the x and y input

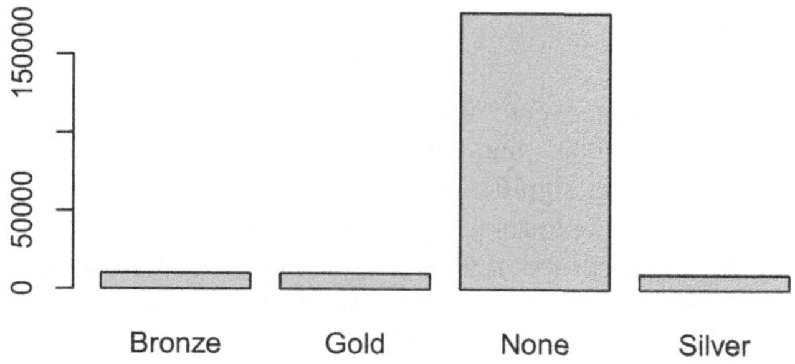

Fig. 2.3 Count of Gold, Silver, and Bronze medals won by Olympic athletes

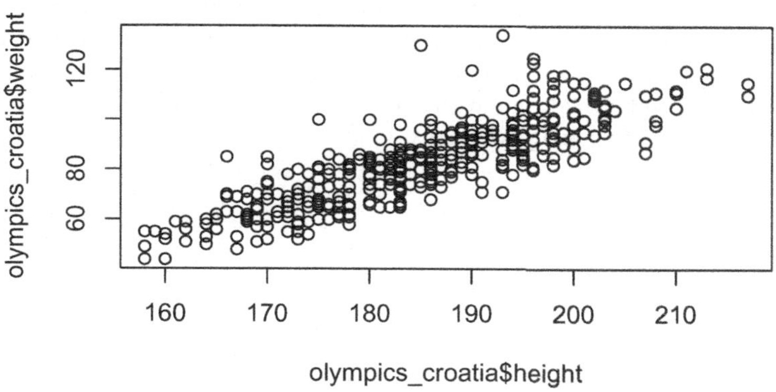

Fig. 2.4 Relationship between height and weight for Croatian Olympic athletes

vectors. For example, let's say we want to create a scatter plot examining the relationship between athlete height and weight in our Olympics dataset for athletes in Croatia. We can do the following:

```
# Subset the Olympics dataset to include only athletes from Croatia
olympics_croatia <- olympics[olympics$team == "Croatia", ]

# Create a scatterplot of athlete height and weight for athletes from
↪  Croatia
plot(olympics_croatia$height, olympics_croatia$weight)
```

As seen in Fig. 2.4, scatter plots are particularly useful when combined with statistical techniques, such as correlation analysis or linear regression. We will look at this more in the linear regression chapter.

2.12 Conclusion

In this chapter, we have covered the basics of R programming, including data manipulation, functions, loops, and conditional statements. We introduced vectorized operations, which can significantly improve the efficiency of your code, and explored basic plotting techniques using base R graphics. This is a lot of ground to cover, and you should feel proud of your progress so far. It is important to remember that learning any programming language takes time, practice, and dedication. Be sure to consistently work through examples, experiment with new techniques, and do not be afraid to ask for help from more experienced users or online communities.

In the next chapter, we will dive into the fundamentals of the "tidyverse." As you will see, the "tidyverse" greatly simplifies data manipulation and visualization tasks in R, making your workflows more efficient and organized. In particular, we will focus on "tidyr" and "dplyr." These packages are arguably the cornerstones of the "tidyverse" and will provide you with a powerful set of tools to efficiently transform and manipulate cultural analytics data.

Chapter 3
Dplyr and TidyR

Abstract This chapter explores the dynamic realm of cultural analytics, highlighting the transformative impact of the tidy data paradigm on data manipulation and analysis. It serves as a practical guide to the "tidyverse," with a particular focus on two of its cornerstone packages, "dplyr" and "tidyr." These tools are instrumental in reshaping complex and messy datasets into an orderly structure suitable for analysis. Through an engaging case study centered on American films from the 1950s to the 2010s, the chapter offers a hands-on demonstration of the power and utility of "tidyverse" functions in real-world scenarios. The case study further explores the merging of disparate data sources through sophisticated joining functions provided by "dplyr." This extends the analytical framework, allowing the reader to combine film attributes with external datasets such as economic indicators.

Keywords Tidyverse · `dplyr` · Data manipulation · Data cleaning · Tidy data · Data transformation · Data wrangling · Statistical data analysis

3.1 Introduction

This chapter introduces the concept of tidy data and the "tidyverse." As we saw in the previous chapter, base R is both powerful and remarkably user-friendly. However, in recent years, the "tidy data" paradigm has grown immensely in popularity. So, what exactly does it mean for data to be "tidy?" We discussed some of this in the introduction, but at its core, the concept of tidy data focuses on a specific structure that makes data easy to manipulate, visualize, and model.

According to Rattenbury et al.:

> There is a common misperception that data analysis is mostly a process of running statistical algorithms on high-performance data engines. In practice, this is just the final step of a longer and more complex process; 50 to 80 percent of an analyst's time is spent wrangling data to get it to the point at which this kind of analysis is possible. (Rattenbury et al. 2017, vii)

Consequently, understanding how to wrangle data into a standard format is essential for effective analysis. But what format should data actually be in? In 2014,

N. Siddiqui, *Cultural Analytics in R: A Tidy Approach*, Use R!,
https://doi.org/10.1007/978-3-031-96618-7_3

statistician Hadley Wickham released a paper outlining his concept of "tidy data" as a potential solution to the difficulty of data wrangling (Wickham 2014). His work draws on research in relational databases, which posits that database normalization is essential to eliminating redundant data and maintaining robustness. While various theories have emerged on how best to do this, the most influential remains Edgar F. Codd's "normal forms" (Codd 1990). Wickham sees "tidy data" as a natural extension of Codd's principles in the realm of statistical data analysis.

In his original paper, he posits data to be tidy if it has the following properties:

1. Each variable forms a column.
2. Each observation forms a row.
3. Each type of observational unit forms a table.

He defines *messy* data as any other arrangement of data. As Wickham contends, "tidy datasets are all alike but every messy dataset is messy in its own way" (Wickham 2014). While messy data has its benefits in certain fields, Wickham argues that it is generally cumbersome and inhibitive to good analytical practices. That is not to say that you cannot analyze messy data. However, it usually requires additional steps of cleaning and restructuring, which not only take additional time but also run the risk of introducing errors into the dataset.

Later, Wickham updated his concept of tidy data to be more specific about what he sees as essential properties of tidy data. Instead of putting as much emphasis on the role of observational units, he adopted a simpler approach that highlights the importance of having a single value in each cell (Wickham et al. 2023):

1. Each variable forms a column.
2. Each observation forms a row.
3. Each value must have its own cell.

This updated form of tidy data makes it even more straightforward for data manipulation and analysis. Along with its philosophy, the tidy data framework has given rise to a powerful set of tools for data manipulation and analysis, collectively known as the "tidyverse." In the "Tidy Tools Manifesto," Wickham notes four general principles that surround the packages that work with tidy data (Wickham et al. 2023):

1. Reuse existing data structures—Although it may not always be ideal, use existing structures, such as rectangular tabular data in a tidy format, whenever possible.
2. Compose simple functions with the pipe—Rather than using complex functions, break down your operations into simpler, more understandable functions. Use the pipe operator (|>) to chain these functions together and name functions after verbs.
3. Embrace functional programming—Avoid using complicated loops and instead apply functions to each component of a list or vector, avoiding side effects like modifying data or global variables within functions.
4. Design for humans—Focus on making APIs and code human-readable, even if it comes at the cost of computational efficiency.

Together, these principles have resulted in a robust ecosystem. In this chapter, we will examine two key packages in the "tidyverse": "dplyr" and "tidyr." "dplyr" offers a set of functions for performing common data manipulation tasks such as filtering, selecting, combining, and mutating data. "tidyr," on the other hand, provides tools for cleaning and reshaping data, making it easier to transform messy data into a tidy format.

We will work with multiple datasets to illustrate the process of tidying data. The first dataset we will use is the MovieLens database, which contains a selection of American films from the 1950s to the 2010s. Compiled by the GroupLens Research team at the University of Minnesota (Harper and Konstan 2015), it contains information on the title, release year, and average rating of each film. Because the duration of each film is not included in this dataset, it was supplemented with data provided by The Movie Database (TMDB), a comprehensive film database created by a community of film enthusiasts.

The second dataset we will use contains genres put together by GroupLens Research and contains information on the genres of each film. We will look at it to illustrate how to join two datasets together to gain more information about the films in our dataset.

The final dataset is a subset from the Gapminder Foundation and contains information about various progress indicator statistics for countries around the world ("About Gapminder | Gapminder," n.d.). These indicators include GDP per capita, life expectancy, and population size, among others. The data is available from 1800 to the present day, providing a comprehensive historical perspective on global development trends. However, we will only use a subset of the data available through the `gapminder` package (Dennis and Daniels n.d.).

Exploring these two datasets provides an opportunity to delve into the intricate evolution of American films, a cornerstone of global culture that has had a significant influence both locally and internationally. Through detailed analysis, we will uncover trends in the average length of American films over the decades. This may reveal changing storytelling techniques, audience attention spans, and industry demands. The study will include examining whether runtime is influenced by the decade of release—reflecting perhaps the historical and technological context of the times—or by the film's genre, suggesting different storytelling conventions and audience expectations within various categories. Additionally, we will analyze the reception of these films by looking at ratings of the films to gain insights into audience preferences.

Finally, we will also consider the dynamic growth of the United States as a nation—its economic development, population changes, and advances in technology—and how these factors have likely played a significant role in shaping its film industry. By correlating developmental milestones with cinematic output, we can quantify the influence of socioeconomic advancements on artistic evolution. This amalgamation of data and culture equips us with the means to understand how a nation's prosperity influences its cultural exports and, in turn, how its cinema affects global cultural currents.

As such, this analytical journey through the history of American films is not merely an exercise in understanding a regional entertainment industry but a reflection on how a nation's cultural narrative can be woven into the global tapestry, reshaping collective cultural understanding and appreciation.

3.2 Installing the Tidyverse

Before we start working with "dplyr" and "tidyr," we need to install the necessary packages. This process is the same as the one we saw when we installed the "magrittr" package in Chap. 1.

Luckily, we can install all of the "tidyverse" at once, which will automatically install the packages we will use in this chapter:

````
```{r}
# Install tidyverse
install.packages("tidyverse")
```
````

Now that we have the "tidyverse" package installed, we need to load it into our R session:

```
# Load tidyverse library
library(tidyverse)
```

Once everything is set up, we can go ahead and begin diving into the data.

3.3 Reading CSV File as Tibble

So far, we have been importing our data by using the `read.csv()` function, which comes from R's built-in "util" package. It is fairly robust and returns the results in a data frame. In the "tidyverse," there is a similar function called `read_csv()`. Unlike the built-in function, the `read_csv()` function stores the results in a tibble. Shortly, we will see what the difference is between a tibble and a data frame. Until then, let's load the data that we have for films from the 1960s to the 1980s:

```
# Load the data
films <- read_csv("data/ch2_films_1960s_1980s.csv")
```

When loading the data, the `read_csv()` call shows the row and column count along with the column names and types. Here we see that we have 4983 films and five columns: one character column ("title") and four double columns ("tmdbId," "rating," "release_year," "runtime"). These columns represent the following information:

tmdbId: The unique identifier for the film in the TMDB database
title: The title of the film
rating: The average rating of the film gathered from the MovieLens database
release_year: The year the film was released
runtime: The duration of the film in minutes gathered from TMDB

We can see that the output of `read_csv()` is different from `read.csv()` because it automatically loads data as a tibble. Let's verify this by using the `class()` function:

```
# Check the class of the object
class(films)
```

```
[1] "spec_tbl_df" "tbl_df"      "tbl"         "data.frame"
```

Tibbles are a modern take on data frames in R, designed specifically to streamline some of their more unwieldy aspects (Müller and Wickham 2023). For instance, one notorious issue of data frames is the automatic conversion of character vectors to factors (categorical variables), an operation that occurs by default when loading data with functions like `read.csv()`. This can be particularly frustrating when you do not want this conversion, as it requires additional work to convert the factors back to characters.

Tibbles also ensure consistency in return types when performing operations. With data frames, certain operations, like subsetting a single column, can sometimes return a vector instead of a data frame. Tibbles avoid this inconsistency by guaranteeing that the result of operations will always be another tibble.

In most contexts, tibbles and data frames can be used interchangeably. However, there are some instances where the choice matters. For example, certain base R functions and packages may not work as expected with tibbles and might require explicit conversion to a data frame. To remedy this, you can explicitly turn a tibble into a data frame using the `as.data.frame()` function and vice versa with `as_tibble()`:

```
# Convert tibble to data frame
films_df <- as.data.frame(films)

# Convert data frame to tibble
films_tibble <- as_tibble(films_df)

# Get the class of the data data.frame
class(films_df)
```

```
[1] "data.frame"
```

As mentioned, in most cases, you can use the two data types interchangeably. However, it's often more convenient to stick with tibbles because of their printing properties and the fact that they are part of the "tidyverse" ecosystem, ensuring seamless compatibility with other "tidyverse" packages and functions. When tibbles are printed, we only see the first ten rows of our data and any columns that can fit on the screen. We also see the type for each of the columns, but the fact that many of the columns do not fit on the screen can make it difficult to tell exactly what is going on:

```
# Print the tibble
films
```

```
# A tibble: 4,983 x 5
   title                          tmdbId rating runtime release_year
   <chr>                           <dbl>  <dbl>   <dbl>        <dbl>
 1 Psycho                            539   4.06     109         1960
 2 One Hundred and One Dalmatians  12230   3.44      82         1961
 3 The Jungle Book                  9325   3.66      78         1967
 4 Mary Poppins                      433   3.73     139         1964
 5 Breakfast at Tiffany's            164   3.82     115         1961
 6 The Birds                         571   3.74     119         1963
 7 The Sword in the Stone           9078   3.56      79         1963
 8 Rosemary's Baby                   805   3.80     138         1968
 9 Planet of the Apes                871   3.62     112         1968
10 The Graduate                    37247   4.02     106         1967
# i 4,973 more rows
```

Another option for viewing tibbles is to use the `glimpse()` function, which provides an output with transposed columns and rows. Ultimately, how you look at your data is a matter of preference, but I usually prefer using `glimpse()` so that is what you will see throughout most of the book when we work with tibbles:

```
# Glimpse the tibble
glimpse(films)
```

```
Rows: 4,983
Columns: 5
$ title        <chr> "Psycho", "One Hundred and One Dalmatians", "The Jungle B~
$ tmdbId       <dbl> 539, 12230, 9325, 433, 164, 571, 9078, 805, 871, 37247, 1~
$ rating       <dbl> 4.062512, 3.444388, 3.656206, 3.728346, 3.815885, 3.73656~
$ runtime      <dbl> 109, 82, 78, 139, 115, 119, 79, 138, 112, 106, 174, 129, ~
$ release_year <dbl> 1960, 1961, 1967, 1964, 1961, 1963, 1963, 1968, 1968, 196~
```

3.3.1 *Cleaning Up NA Values*

While this particular dataset does not contain any missing values, it is common to encounter them in real-world datasets. Missing values can be problematic for data analysis, as they can lead to errors in calculations and visualizations. Therefore, it is essential to identify and handle missing values appropriately.

In R, missing or unavailable pieces of data are usually coded using the notation `NA`, which stands for "Not Available." The convention of using `NA` is deeply embedded within R's framework and enables the handling of incomplete data through a variety of established procedures. These procedures include the capacity to exclude missing values in analytical functions or to apply sophisticated imputation techniques that estimate and substitute missing data with plausible values.

In the data folder, there is a file called "na_films.csv" that contains the same data as our original film dataset, but with some titles listed as "missing." Let's load this data to see how many NA values are present:

```
# Load the data with missing values
na_films <- read_csv("data/ch2_na_films.csv")

# Glimpse the data
glimpse(na_films)
```

```
Rows: 4,983
Columns: 5
$ title        <chr> "Psycho", "One Hundred and One Dalmatians", "The Jungle B~
$ tmdbId       <dbl> 539, 12230, 9325, 433, 164, 571, 9078, 805, 871, 37247, 1~
$ rating       <dbl> 4.062512, 3.444388, 3.656206, 3.728346, 3.815885, 3.73656~
$ runtime      <dbl> 109, 82, 78, 139, 115, 119, 79, 138, 112, 106, 174, 129, ~
$ release_year <dbl> 1960, 1961, 1967, 1964, 1961, 1963, 1963, 1968, 1968, 196~
```

We see that we have the same amount of films represented by the rows, but if we sum the number of missing values in the rating column, we find that there are numerous missing values:

```
# Find the number of missing values
sum(na_films$title == "missing")
```

```
[1] 507
```

There are 507 missing titles that now need to be converted to `NA` values. Rather than doing this manually, we can use the `"na.strings"` argument in the `read_csv()` function. This argument allows us to specify which values should be treated as missing values. In this case, we will set `"na.strings"` to "missing" to convert the "missing" values to `NA`:

```
# Load the data with missing values
na_films <- read_csv("data/ch2_na_films.csv", na = "missing")

# Glimpse the data
glimpse(na_films)
```

```
Rows: 4,983
Columns: 5
$ title        <chr> "Psycho", "One Hundred and One Dalmatians", "The Jungle B~
```

```
$ tmdbId       <dbl> 539, 12230, 9325, 433, 164, 571, 9078, 805, 871, 37247, 1~
$ rating       <dbl> 4.062512, 3.444388, 3.656206, 3.728346, 3.815885, 3.73656~
$ runtime      <dbl> 109, 82, 78, 139, 115, 119, 79, 138, 112, 106, 174, 129, ~
$ release_year <dbl> 1960, 1961, 1967, 1964, 1961, 1963, 1963, 1968, 1968, 196~
```

Let's now count up the number of NA values in the title column to ensure that the conversion was successful:

```
# Find the number of missing values

sum(is.na(na_films$title))
```

```
[1] 507
```

We see that we get the same number of missing values as before. However, by using R's built-in `NA` values, we were able to use built-in functions like `is.na()` to identify and handle issues in our data.

3.4 Dplyr and the Five Verbs

We are now ready to begin examining our dataset. In this domain, you'll quickly discover that even though every dataset is unique, the fundamental tasks employed in their analysis are often similar, if not identical. These tasks, performed repetitively, form the bedrock of data examination and manipulation. For instance, irrespective of your dataset, you'll frequently find yourself needing to select or remove specific columns that are relevant to your inquiry or filter rows based on specific conditions. Previously, we explored how to undertake these tasks using base R, but we'll now delve into a different approach via the "dplyr" package.

"dplyr" is a comprehensive and efficient toolkit designed for data manipulation in R (Wickham et al. 2023). It provides a consistent and accessible set of "verbs"—functions that perform specific actions—which simplify the most common data manipulation challenges. These verbs, such as `select()`, `filter()`, and `mutate()`, form the cornerstone of the package's functionality. As we will see, we can chain a series of these calls together through the pipe operator to create an end-to-end data manipulation pipeline.

3.4.1 Filtering Out Rows

Often, you need to select a group of rows based on some condition. The equivalent function for filtering out rows in a dataset using the "tidyverse" is aptly called `filter()`. This function takes two main arguments: the dataset and the condition that rows need to meet to be included in the result.

For the sake of example, let's remove films released in the year 1960 from our dataset. We can do this by using the `filter()` function and specifying the condition that the "release_year" column should not be equal to 1960:

```
# Filter out films released in 1960
films_filtered <- films |>
  filter(release_year != 1960)

# Glimpse the filtered data
glimpse(films_filtered)
```

```
Rows: 4,886
Columns: 5
$ title        <chr> "One Hundred and One Dalmatians", "The Jungle Book", "Mar~
$ tmdbId       <dbl> 12230, 9325, 433, 164, 571, 9078, 805, 871, 37247, 15121,~
$ rating       <dbl> 3.444388, 3.656206, 3.728346, 3.815885, 3.736569, 3.55849~
$ runtime      <dbl> 82, 78, 139, 115, 119, 79, 138, 112, 106, 174, 129, 173, ~
$ release_year <dbl> 1961, 1967, 1964, 1961, 1963, 1963, 1968, 1968, 1967, 196~
```

We see that the amount of films we have has gone down to 4886 from 4983. This is because we have removed the 97 films that were released in 1960. We can also see that the "release_year" column no longer contains the value 1960.

What if we were also interested in removing any film that did not have a rating above 3? We can combine the filter for the "release_year" and the "rating" by using the & operator as below:

```
# Filter out films released in 1980 and with a rating below 3
films_filtered <- films |>
  filter(release_year != 1960 & rating > 3)

# Glimpse the filtered data
glimpse(films_filtered)
```

```
Rows: 2,207
Columns: 5
$ title        <chr> "One Hundred and One Dalmatians", "The Jungle Book", "Mar~
$ tmdbId       <dbl> 12230, 9325, 433, 164, 571, 9078, 805, 871, 37247, 15121,~
$ rating       <dbl> 3.444388, 3.656206, 3.728346, 3.815885, 3.736569, 3.55849~
$ runtime      <dbl> 82, 78, 139, 115, 119, 79, 138, 112, 106, 174, 129, 173, ~
$ release_year <dbl> 1961, 1967, 1964, 1961, 1963, 1963, 1968, 1968, 1967, 196~
```

While combining the conditions with the & operator is useful, it can become cumbersome when dealing with multiple conditions. In such cases, you can use the |> operator to chain multiple `filter()` calls together. You will see that the "tidyverse" makes heavy use of this operator, so if you do not understand what it does, now is a good time to review the previous chapter. Here is how we can rewrite the filter command described earlier using pipes:

```
# Filter out films released in 1980 and with a rating below 3
films_1980s <- films |>
  filter(release_year != 1980) |>
  filter(rating > 3)

# Glimpse the filtered data
glimpse(films_filtered)
```

```
Rows: 2,207
Columns: 5
$ title        <chr> "One Hundred and One Dalmatians", "The Jungle Book", "Mar~
$ tmdbId       <dbl> 12230, 9325, 433, 164, 571, 9078, 805, 871, 37247, 15121,~
$ rating       <dbl> 3.444388, 3.656206, 3.728346, 3.815885, 3.736569, 3.55849~
$ runtime      <dbl> 82, 78, 139, 115, 119, 79, 138, 112, 106, 174, 129, 173, ~
$ release_year <dbl> 1961, 1967, 1964, 1961, 1963, 1963, 1968, 1968, 1967, 196~
```

Remember that one of the ways we can read the pipe operator is by using the words "and then." So, in the code above, we are saying, "take the 'films_1980s' dataset and then filter out the films released in 1980 and then filter out the films with a rating below 3."

3.4.2 Selecting Specific Columns

In some cases, you may only be interested in working with certain columns in your dataset. While there is no harm in keeping extraneous columns, it can make your analysis more difficult and slow down your code, especially when working with large datasets. This is where the `select()` function comes in. If `filter()` allows you to work only with the rows you want, `select()` does the same thing for columns by allowing you to choose which columns you want to keep or remove.

Let's say we no longer want to focus on the runtime of the films. We can use the `select()` function to choose the other five columns in our dataset:

```
# Select specific columns
films_selected <- films |>
  select(tmdbId, title, rating, release_year, runtime)

# Glimpse the selected data
glimpse(films_selected)
```

```
Rows: 4,983
Columns: 5
$ tmdbId       <dbl> 539, 12230, 9325, 433, 164, 571, 9078, 805, 871, 37247, 1~
$ title        <chr> "Psycho", "One Hundred and One Dalmatians", "The Jungle B~
$ rating       <dbl> 4.062512, 3.444388, 3.656206, 3.728346, 3.815885, 3.73656~
$ release_year <dbl> 1960, 1961, 1967, 1964, 1961, 1963, 1963, 1968, 1968, 196~
$ runtime      <dbl> 109, 82, 78, 139, 115, 119, 79, 138, 112, 106, 174, 129, ~
```

It can be tedious to type out the names of all your columns just to get rid of a few. In these scenarios, you can use - in front of the column names to signal that you want to remove a column. The following code will provide the same result as above:

```
# Select specific columns
films_selected <- films |>
  select(-runtime)

# Glimpse the selected data
glimpse(films_selected)
```

```
Rows: 4,983
Columns: 4
$ title        <chr> "Psycho", "One Hundred and One Dalmatians", "The Jungle B~
$ tmdbId       <dbl> 539, 12230, 9325, 433, 164, 571, 9078, 805, 871, 37247, 1~
$ rating       <dbl> 4.062512, 3.444388, 3.656206, 3.728346, 3.815885, 3.73656~
$ release_year <dbl> 1960, 1961, 1967, 1964, 1961, 1963, 1963, 1968, 1968, 196~
```

3.4.3 Arranging Our Data

There are many instances in data analysis where you might find it beneficial or even necessary to reorganize the sequence of rows in your dataset. This could be due to various reasons, such as wanting to view your data from a distinct perspective, aiming to identify certain patterns more effortlessly, or even as a prerequisite for certain types of analysis. Such reordering is typically performed based on one or multiple variables within your dataset.

The `arrange()` function incorporated within "dplyr" is designed to facilitate this task. It receives, as its first argument, the dataset you intend to rearrange. Following this, you can specify the variables by which you want to sort the dataset as additional arguments. Let's apply this to our dataset with our film dataset. Suppose we are interested in viewing our data in terms of the "release_year." We can use the `arrange()` function to sort our dataset based on this variable:

```
# Arrange the data by title
films_arranged <- films |>
  arrange(release_year)

# Glimpse the arranged data
glimpse(films_arranged)
```

```
Rows: 4,983
Columns: 5
$ title        <chr> "Psycho", "The Apartment", "Spartacus", "The Magnificent ~
$ tmdbId       <dbl> 539, 284, 967, 966, 2134, 1908, 24452, 11209, 299, 18444,~
$ rating       <dbl> 4.062512, 4.028898, 3.826786, 3.898370, 3.699530, 4.05269~
$ runtime      <dbl> 109, 125, 197, 127, 103, 128, 72, 167, 127, 126, 79, 208,~
$ release_year <dbl> 1960, 1960, 1960, 1960, 1960, 1960, 1960, 1960, 1960, 196~
```

As you can see, the `arrange()` function defaults to ascending order. If you want to sort the dataset in descending order, you can use the `desc()` function within the `arrange()` function. For example, if we wanted to sort our dataset based on the year of release in descending order, we would do the following:

```
# Arrange the data by release year in descending order
films_arranged <- films |>
  arrange(desc(release_year))

# Glimpse the arranged data
glimpse(films_arranged)
```

```
Rows: 4,983
Columns: 5
$ title        <chr> "Back to the Future Part II", "Dead Poets Society", "Indi~
$ tmdbId       <dbl> 165, 207, 89, 10144, 268, 2978, 639, 942, 9354, 2756, 949~
$ rating       <dbl> 3.593533, 3.938424, 3.986865, 3.535704, 3.409920, 3.05033~
$ runtime      <dbl> 108, 128, 127, 83, 126, 108, 96, 114, 93, 140, 93, 97, 10~
$ release_year <dbl> 1989, 1989, 1989, 1989, 1989, 1989, 1989, 1989, 1989, 198~
```

You can also sort the dataset based on multiple variables. In this case, the function will sort by the first variable and then by the second one. For instance, if we wanted to sort by year of release and then rating in ascending order, we could do the following:

```
# Arrange the data by release year and title
films_arranged <- films |>
  arrange(release_year, rating)

# Glimpse the arranged data
glimpse(films_arranged)
```

```
Rows: 4,983
Columns: 5
$ title        <chr> "13 Fighting Men", "Sex Kittens Go to College", "Dinosaur~
$ tmdbId       <dbl> 359195, 116373, 55663, 88376, 89751, 89763, 31286, 18977,~
$ rating       <dbl> 0.500000, 1.500000, 1.750000, 1.777778, 1.833333, 1.87500~
$ runtime      <dbl> 69, 94, 85, 85, 78, 78, 71, 82, 112, 74, 94, 62, 75, 77, ~
$ release_year <dbl> 1960, 1960, 1960, 1960, 1960, 1960, 1960, 1960, 1960, 196~
```

In many cases, you may need to create new columns in your dataset based on existing columns. This process is known as "mutating" and is a common operation in data analysis. The "dplyr" package provides the `mutate()` function to facilitate this task. The function takes the dataset as its first argument and then allows you to create new columns based on the existing columns.

For example, let's say we want a new column that has "runtime_in_hours" instead of "runtime" in minutes. We can create this new column by dividing the "runtime" column by 60:

```
# Mutate the data to create a new column
films_mutated <- films |>
  mutate(runtime_in_hours = runtime / 60)

# Glimpse the mutated data
glimpse(films_mutated)
```

```
Rows: 4,983
Columns: 6
$ title            <chr> "Psycho", "One Hundred and One Dalmatians", "The Jung~
$ tmdbId           <dbl> 539, 12230, 9325, 433, 164, 571, 9078, 805, 871, 3724~
$ rating           <dbl> 4.062512, 3.444388, 3.656206, 3.728346, 3.815885, 3.7~
$ runtime          <dbl> 109, 82, 78, 139, 115, 119, 79, 138, 112, 106, 174, 1~
$ release_year     <dbl> 1960, 1961, 1967, 1964, 1961, 1963, 1963, 1968, 1968,~
$ runtime_in_hours <dbl> 1.816667, 1.366667, 1.300000, 2.316667, 1.916667, 1.9~
```

Working with this data for many different years may make it difficult to tease out broader trends in film production. Let's create a new column that has the decade for the film. In other words, if a film's release was from 1960 to 1969, the "decade" column would say 1960s, and if it was from 1970 to 1979, it would say 1970s. We can use nested `ifelse()` functions to do this, but a better way is through the `case_when()` function. The `case_when()` function takes a series of conditionals and returns a specific value depending on the result of that conditional. The relationship between the conditional and its result is signified by the tilde (~) operator. You may recognize this as a switch statement if you have worked in other programming languages. We will store this data back into the film dataset so that we can use it later:

```
# Mutate the data to create a new column using case_when
films <- films |>
  mutate(
    decade = case_when(
      release_year >= 1950 & release_year <= 1959 ~ "1950s",
      release_year >= 1960 & release_year <= 1969 ~ "1960s",
      release_year >= 1970 & release_year <= 1979 ~ "1970s",
      release_year >= 1980 & release_year <= 1989 ~ "1980s",
      TRUE ~ "unknown"
    )
  )

# Glimpse the mutated data
glimpse(films)
```

```
Rows: 4,983
Columns: 6
$ title        <chr> "Psycho", "One Hundred and One Dalmatians", "The Jungle B~
$ tmdbId       <dbl> 539, 12230, 9325, 433, 164, 571, 9078, 805, 871, 37247, 1~
$ rating       <dbl> 4.062512, 3.444388, 3.656206, 3.728346, 3.815885, 3.73656~
$ runtime      <dbl> 109, 82, 78, 139, 115, 119, 79, 138, 112, 106, 174, 129, ~
$ release_year <dbl> 1960, 1961, 1967, 1964, 1961, 1963, 1963, 1968, 1968, 196~
$ decade       <chr> "1960s", "1960s", "1960s", "1960s", "1960s", "1960s", "19~
```

The last statement may need to be clarified. It is a catch-all for any values not picked up by the first four and can be particularly useful for catching errors. Since we have presumably accounted for all of the decades in our dataset, those left are simply errors in our dataset. We will place "unknown" as the value for them. While this particular dataset does not have any errors, it is good practice to include a catch-all for any unexpected values that may arise in the future.

Sometimes, you only want the new column you calculated. For this, you can use the `transmute()` function. This function works similarly to `mutate()`, but only keeps the new column(s) you create. For the sake of example, let's create a tibble where we only keep the runtime hours. We will not store this result and will only print it to the console:

```
# Transmute the data to create a new column
films_transmuted <- films |>
  transmute(runtime_in_hours = runtime / 60)

# Print the transmuted data
films_transmuted
```

```
# A tibble: 4,983 x 1
   runtime_in_hours
              <dbl>
 1             1.82
 2             1.37
 3             1.3
 4             2.32
 5             1.92
 6             1.98
 7             1.32
 8             2.3
 9             1.87
10             1.77
# i 4,973 more rows
```

3.4.4 Summarizing and Grouping Data

Analyzing the key statistical attributes of a dataset is vital to comprehending the properties and distributions of your data. Tasks like determining central tendencies, variability, or extremes often serve as the first step to unlocking essential insights. The `summarize()` function from "dplyr" provides a handy mechanism to compute such summary statistics conveniently.

This function accepts your dataset as the first argument, followed by the specific statistical measures you wish to calculate. These could include the minimum, maximum, mean, median, standard deviation, or any other relevant metric. After using `summarize()`, you will receive a new dataset containing the summary values you requested. Let's get the average runtime of the films along with the average rating. We can calculate this using the following code:

```
# Summarize the data
films_summarized <- films |>
  summarize(
    mean_runtime = mean(runtime),
    mean_rating = mean(rating),
  )

# Print the summarized data
films_summarized
```

```
# A tibble: 1 x 2
  mean_runtime mean_rating
         <dbl>       <dbl>
1         99.3        2.90
```

We see that the minimum runtime for our dataset is 99.3 minutes, while the average rating is 2.90. Together, these statistics can help us understand the general trends in our dataset. However, it is important to remember that these statistics are only as good as the data they are based on. If there are missing values or errors in the data, the summary statistics may not be accurate. Therefore, it is crucial to ensure that your data is clean and reliable before performing any analysis.

You can combine `summarize()` with the `group_by()` function to calculate summary statistics for specific groups within your dataset. The `group_by()` function divides your dataset by one or more variables, allowing you to perform calculations on each group separately. For instance, let's calculate the average runtime for films released in each of the decades. We can do this by grouping the data by the "decade" column we created earlier and then summarizing the runtime for each group:

```
# Group the data by decade half and summarize the runtime
films_1980s_grouped <- films |>
  group_by(decade) |>
  summarize(
    mean_runtime = mean(runtime)
  )

# Print the grouped data
films_1980s_grouped
```

```
# A tibble: 3 x 2
  decade mean_runtime
  <chr>         <dbl>
1 1960s         102. 
2 1970s          97.9
3 1980s          98.9
```

We see that the mean runtime for films released in the 1960s is 102 minutes. In comparison, the 1970s and 1980s have similar runtimes with 97.9 and 98.9 respectively. This small difference suggests minimal variation in average film runtimes between the two periods. We could explore this more by performing a statistical test to determine if this difference is significant or not. In addition, it may be helpful to group by "genre" or other variables to uncover more nuanced patterns in film runtimes.

You will find that you often need to know how many rows fall into a certain subgroup or have a specific attribute value in your dataset. For this, we can use the `n()` function within `summarize()` to find out the number of rows within each group. For example, if we wanted to know the number of films in each decade, we could do the following:

```
# Group the data by decade half and summarize the number of films
films_1980s_grouped <- films |>
  group_by(decade) |>
  summarize(
    n_films = n()
  )

# Print the grouped data
films_1980s_grouped
```

```
# A tibble: 3 x 2
  decade n_films
  <chr>    <int>
1 1960s     1107
2 1970s     1688
3 1980s     2188
```

We see that by the 1980s, our dataset contains almost double the amount of films from the 1960s. This could be due to a number of factors, such as increased film production during that decade or simply a reflection of the particular catalog from which the dataset was sourced.

3.4.5 Grouping and Ungrouping Variables

In the process of data analysis using R, particularly with the "dplyr" package, it's crucial to understand the implications of employing the `summarize()` function following the `group_by()` function. When you use `summarize()`, it aggregates the data based on the specified grouping and calculates the summary statistics or other results only for the variables included in the summarize expression. Consequently, the resulting dataset contains only the grouping variables and the summarized variables. This means that other columns from the original dataset are not retained in the output. Essentially, `summarize()` produces a new, more condensed data frame where the focus is on the aggregated values, often losing the individual row-level detail that was present in the original dataset.

This behavior can sometimes pose a limitation, especially if you need to perform group-wise calculations without losing access to the full dataset. For example, you might want to calculate the average of a particular variable within subgroups of your data and then add this average as a new column to the original dataset for further analysis. To manage such a scenario, it's necessary to use a different approach that allows you to maintain the granular information.

For instance, let's say we want to calculate the difference between a film's runtime and the average for that decade. The first thing we need to do is group by the decade and calculate the mean for that decade. We will put that in a new column entitled "decade_average":

```
# Group the data by decade and calculate the mean runtime
films_grouped <- films |>
  group_by(decade) |>
  mutate(decade_average = mean(runtime))
```

If we use `glimpse()` to look at our data, we can see that we have a new column called "decade_average" that contains the average runtime for each decade:

```
# Glimpse the data
glimpse(films_grouped)
```

```
Rows: 4,983
Columns: 7
Groups: decade [3]
$ title          <chr> "Psycho", "One Hundred and One Dalmatians", "The Jungle~
$ tmdbId         <dbl> 539, 12230, 9325, 433, 164, 571, 9078, 805, 871, 37247,~
$ rating         <dbl> 4.062512, 3.444388, 3.656206, 3.728346, 3.815885, 3.736~
$ runtime        <dbl> 109, 82, 78, 139, 115, 119, 79, 138, 112, 106, 174, 129~
$ release_year   <dbl> 1960, 1961, 1967, 1964, 1961, 1963, 1963, 1968, 1968, 1~
$ decade         <chr> "1960s", "1960s", "1960s", "1960s", "1960s", "1960s", "~
$ decade_average <dbl> 102.1653, 102.1653, 102.1653, 102.1653, 102.1653, 102.1~
```

Let's see what happens, however, when we call the `class()` function:

```
# Check the class of the object
class(films_grouped)
```

```
[1] "grouped_df" "tbl_df"     "tbl"        "data.frame"
```

We see that we now have a grouped data frame. This happens because once we have performed operations on a grouped data frame in R, the data remains partitioned into the groups specified by the `group_by()` function. This grouped structure can restrict subsequent operations to within those groups, which can be useful for certain calculations but might interfere with other types of analysis or manipulations that require considering the dataset as a whole. To release the data from the constraints of the grouping, we use the `ungroup()` function. Applying `ungroup()` is akin to resetting the data frame to its original state where no grouping is applied. Post-ungrouping, the data frame is no longer segmented by the grouping variables, and any operation you perform thereafter acts on the dataset in its entirety.

Let us go ahead and see how this all works in practice by recreating the "decade_average_runtime" column and another one called "decade_runtime_deviation," which shows the difference between the runtime of a film and the average runtime for that decade. Because the difference from the runtime could be either positive or negative, we will use the `abs()` function to find the absolute value of the difference between the runtime and the decade average:

```
# Group the data by decade and calculate the mean runtime. Use ungroup to
↪   remove the grouping.

films <- films |>
  group_by(decade) |>
  mutate(
    decade_average_runtime = mean(runtime),
  ) |>
  ungroup() |>
  mutate(
    decade_runtime_deviation = abs(runtime - decade_average_runtime)
  )

# Glimpse the data
glimpse(films)
```

```
Rows: 4,983
Columns: 8
$ title                    <chr> "Psycho", "One Hundred and One Dalmatians", "~
$ tmdbId                   <dbl> 539, 12230, 9325, 433, 164, 571, 9078, 805, 8~
$ rating                   <dbl> 4.062512, 3.444388, 3.656206, 3.728346, 3.815~
$ runtime                  <dbl> 109, 82, 78, 139, 115, 119, 79, 138, 112, 106~
$ release_year             <dbl> 1960, 1961, 1967, 1964, 1961, 1963, 1963, 196~
$ decade                   <chr> "1960s", "1960s", "1960s", "1960s", "1960s", ~
$ decade_average_runtime   <dbl> 102.1653, 102.1653, 102.1653, 102.1653, 102.1~
$ decade_runtime_deviation <dbl> 6.834688, 20.165312, 24.165312, 36.834688, 12~
```

Now that we have our new column, we can arrange by the highest value to see the films with the greatest runtime difference from the "decade_average_runtime":

```
# Arrange the data by the highest runtime deviation
films |>
  arrange(desc(decade_runtime_deviation))
```

```
# A tibble: 4,983 x 8
   title          tmdbId rating runtime release_year decade decade_average_runtime
   <chr>           <dbl>  <dbl>   <dbl>        <dbl> <chr>                   <dbl>
 1 Near Death      93540   4.67     358         1989 1980s                    98.9
 2 Sleep          256222   2.1      321         1964 1960s                   102. 
 3 The Art of ~    94721   4        278         1965 1960s                   102. 
 4 Elvis and Me    23178   2.8      240         1988 1980s                    98.9
 5 A Woman Cal~   126031   3.2      240         1982 1980s                    98.9
 6 Inside the ~    99357   3.17     240         1982 1980s                    98.9
 7 Earth Star ~    42376   2.83     240         1988 1980s                    98.9
 8 The Iceman ~    41461   3.44     239         1973 1970s                    97.9
 9 Renaldo and~   168595   4        235         1978 1970s                    97.9
10 Woodstock        9459   3.90     225         1970 1970s                    97.9
# i 4,973 more rows
# i 1 more variable: decade_runtime_deviation <dbl>
```

The documentary *Near Death* stands out as the film with the most significant deviation in runtime from its respective decade. With a staggering runtime of 358 minutes, it exceeds the average 1980's film duration by 259 minutes. Similarly, the documentary *Sleep* from the 1960s exhibits an extended runtime of 219 minutes beyond the era's average film length. This notable trend raises the question: why do documentaries often feature such extensive runtimes?

Documentaries frequently delve into complex and multifaceted subjects, demanding considerable screen time to explore thoroughly. Additionally, documentaries are not bound by the same commercial constraints that typically govern mainstream cinema, allowing filmmakers greater freedom to extend their narratives without editing for brevity. This creative liberty enables a more comprehensive examination of their topics.

3.5 Combining and Joining Tables

Having explored the foundational verbs of the "dplyr" package and the process of structuring our data into a tidy format, we are now poised to delve into the realm of relational data. In the landscape of data analysis, it is quite commonplace to encounter datasets that possess inherent relationships. These relationships often manifest through commonalities or references that link one table to another, thereby forming a network of interrelated information.

For instance, consider a scenario where one has access to a dataset enumerating details about films—such as their titles, release dates, and genre classifications—

and a separate dataset cataloging information pertinent to directors, including their names, filmographies, and career achievements. In this case, a natural relationship exists between the two datasets: films are directed by directors. To analyze such interconnected data effectively, we need to merge these datasets in a meaningful manner that accurately represents the relationships between them.

The “dplyr” package comes equipped with functions specifically designed to handle the merging and joining of related tables. By leveraging these functions, we can enrich our analysis with a multifaceted view of the data that considers the complete picture formed by the aggregate datasets. There are many ways to combine tables, but there are two key operations: unions and joins.

3.5.1 Combining Rows with Unions

Unions are a type of data operation that involves merging two or more tables by stacking them vertically, effectively appending the rows of one table to another. This operation is crucial when we aim to consolidate similar datasets to achieve a comprehensive perspective. While our current dataset contains films from the 1960s to the 1980s, we also have data for films from the 1990s to the 2010s. To merge these two datasets, we can use the `bind_rows()` function from the “dplyr” package.

Let’s look at how this works in practice by loading in all of our datasets and then combining them into one. As before, we will use the `read_csv()` function to load in the additional data:

```
# Load the data for films from the 1990s to the 2010s
films_1990s <- read_csv("data/ch2_films_1990s.csv")
films_2000s <- read_csv("data/ch2_films_2000s.csv")
films_2010s <- read_csv("data/ch2_films_2010s.csv")
```

Now that we have read our data, we combine it into one tibble. To do this, we can use the `bind_rows()` function. This function takes two or more tibbles as its arguments and stacks them on each other. The resulting tibble combines the rows from each input tibble. Let us go ahead and combine our datasets:

```
# Combine the datasets
films_1990s_2010s <- bind_rows(films_1990s, films_2000s, films_2010s)
```

We can look at the combined results using `glimpse()`:

```
# Glimpse the combined data
glimpse(films)
```

```
Rows: 4,983
Columns: 8
$ title                   <chr> "Psycho", "One Hundred and One Dalmatians", "~
```

```
$ tmdbId                    <dbl> 539, 12230, 9325, 433, 164, 571, 9078, 805, 8~
$ rating                    <dbl> 4.062512, 3.444388, 3.656206, 3.728346, 3.815~
$ runtime                   <dbl> 109, 82, 78, 139, 115, 119, 79, 138, 112, 106~
$ release_year              <dbl> 1960, 1961, 1967, 1964, 1961, 1963, 1963, 196~
$ decade                    <chr> "1960s", "1960s", "1960s", "1960s", "1960s", ~
$ decade_average_runtime    <dbl> 102.1653, 102.1653, 102.1653, 102.1653, 102.1~
$ decade_runtime_deviation <dbl> 6.834688, 20.165312, 24.165312, 36.834688, 12~
```

Let's also create the columns that we created earlier for the 1960's to the 1980's dataset. This will allow us to compare the runtime and rating of films across the different decades. Because "dplyr" allows us to chain operations together, we can do this in one step:

```
# Create the decade column and calculate the average runtime. Then,
↪  ungroup the data and calculate the runtime deviation.

films_1990s_2010s <- films_1990s_2010s |>
  mutate(runtime_hours=runtime/60) |>
  mutate(
    decade = case_when(
      release_year >= 1990 & release_year <= 1999 ~ "1990s",
      release_year >= 2000 & release_year <= 2009 ~ "2000s",
      release_year >= 2010 & release_year <= 2019 ~ "2010s",
      TRUE ~ "unknown"
    )
  ) |>
  group_by(decade) |>
  mutate(
    decade_average_runtime = mean(runtime),
  ) |>
  ungroup() |>
  mutate(
    decade_runtime_deviation = abs(runtime - decade_average_runtime)
      )
```

Now that our new films tibble has been adjusted with the new decades and additional steps, we can combine with our previous dataset to develop a comprehensive overview of the American film industry from the 1960s to the 2010s—at least as it is related to our dataset:

```
# Combine the datasets
films <- bind_rows(films, films_1990s_2010s)

# Glimpse the combined data
glimpse(films)
```

```
Rows: 20,835
Columns: 9
$ title                     <chr> "Psycho", "One Hundred and One Dalmatians", "~
$ tmdbId                    <dbl> 539, 12230, 9325, 433, 164, 571, 9078, 805, 8~
$ rating                    <dbl> 4.062512, 3.444388, 3.656206, 3.728346, 3.815~
```

```
$ runtime                  <dbl> 109, 82, 78, 139, 115, 119, 79, 138, 112, 106~
$ release_year             <dbl> 1960, 1961, 1967, 1964, 1961, 1963, 1963, 196~
$ decade                   <chr> "1960s", "1960s", "1960s", "1960s", "1960s", ~
$ decade_average_runtime   <dbl> 102.1653, 102.1653, 102.1653, 102.1653, 102.1~
$ decade_runtime_deviation <dbl> 6.834688, 20.165312, 24.165312, 36.834688, 12~
$ runtime_hours            <dbl> NA, NA, NA, NA, NA, NA, NA, NA, NA, NA, NA, N~
```

3.5.2 *Combining Columns with Joins*

3.6 Loading in Film Genre Data

Right now, our data consists of data about the films. However, we may also be interested in looking at the genres for each film. To do this, we can use the concept of joins in merging our existing tibble containing film data with another tibble containing genre data. The genre data is stored in a separate file called "film_genres.csv" in the data folder. Let's load this data into a tibble and take a look at it:

```
# Load the genre data
genres <- read_csv("data/ch2_films_genres.csv")

# Glimpse the genre data
glimpse(genres)
```

```
Rows: 171,962
Columns: 2
$ tmdbId <dbl> 24428, 293660, 299536, 550, 118340, 680, 13, 1726, 68718, 278, ~
$ genres <chr> "Science Fiction, Action, Adventure", "Action, Adventure, Comed~
```

The genre data contains two columns: "tmdbId" and "genre." The "tmdbId" column is the unique identifier for each film in the TMDB database, while the "genre" column contains a list of genres for each film. The genres are separated by a comma (`,`) character. This format is common in datasets where a single column contains multiple values separated by a delimiter. Depending on the analysis you want to perform, you may need to split these values into separate columns. However, for now, we will keep the genres as they are to focus on the merging process.

Joins require a common thread, a "key" that exists in both datasets, serving as the connective tissue that links related records together. This "key" is often a unique identifier that appears in both tables, such as a film's ID or title. The presence of a key ensures that the merge operation respects the relational context, pairing each film's data with its corresponding rating in a coherent and accurate manner. By utilizing different types of joins, each suited for specific scenarios and outcomes, we can tailor the merge process to fit our analysis goals.

The nature of the relationship between the datasets guides us in choosing among the five main types of joins, each supported by its respective function in the "dplyr" package:

Left Outer Join—Using a key, the entries in the two corresponding tables are joined. The left table entry will still be preserved if an entry does not have a corresponding match. We can do a left outer join using the `left_join()` function.

Right Outer Join—Using a key, the entries in the corresponding tables are joined. The right table entry will still be preserved if an entry does not have a corresponding match. We can do a right outer join using the `right_join()` function.

Inner Join—Using a key, the entries in the corresponding tables are joined. If an entry does not have a corresponding match, it will not be preserved. We can do an inner join using the `inner_join()` function.

Full Outer Join—Using a key, the entries in the two tables that correspond are joined. All entries that do not have a corresponding match will be preserved. We can do a full outer join (often shortened to full join) using the `full_join()` function.

Anti-join—Using a key, the entries in the two tables that do not correspond are joined. All entries that have a corresponding match will not be preserved. We can do an anti-join using the `anti_join()` function.

For each of these joins, the first argument is the dataset you want to merge with, and the second "by" argument is set to the column key that you want to merge on.

For our purposes, we will use an inner join to merge the genre data with our film data. This will allow us to keep only the films that have a genre associated with them. However, the other joins work in a similar manner, so you can use them as needed for your analysis:

```
# Join the films data with the genre data
films<- films |>
  inner_join(genres, by = "tmdbId")

# Glimpse the joined data
glimpse(films)
```

```
Rows: 20,857
Columns: 10
$ title                      <chr> "Psycho", "One Hundred and One Dalmatians", "~
$ tmdbId                     <dbl> 539, 12230, 9325, 433, 164, 571, 9078, 805, 8~
$ rating                     <dbl> 4.062512, 3.444388, 3.656206, 3.728346, 3.815~
$ runtime                    <dbl> 109, 82, 78, 139, 115, 119, 79, 138, 112, 106~
$ release_year               <dbl> 1960, 1961, 1967, 1964, 1961, 1963, 1963, 196~
$ decade                     <chr> "1960s", "1960s", "1960s", "1960s", "1960s", ~
$ decade_average_runtime     <dbl> 102.1653, 102.1653, 102.1653, 102.1653, 102.1~
$ decade_runtime_deviation <dbl> 6.834688, 20.165312, 24.165312, 36.834688, 12~
$ runtime_hours              <dbl> NA, NA, NA, NA, NA, NA, NA, NA, NA, NA, NA, N~
$ genres                     <chr> "Horror, Drama, Thriller", "Adventure, Animat~
```

3.7 Data Transformation and Cleanup

The datasets we have looked at so far have already been in a tidy format. However, you will frequently find "messy" or "untidy" datasets in the wild. It is important to note that data in this format is not necessarily "wrong"; rather, it reflects a structure that may serve specific functions or adhere to standards within particular domains or applications. When engaging in cultural analysis, you will likely be working with data from various sources, so knowing the reasons for its original structure is important. Still, while keeping data in the original format allows us to retain its context and authenticity, reshaping it provides two fundamental benefits: first, we can utilize the powerful tools in the tidyverse to manipulate and analyze the data effectively; and second, reshaping the data into a consistent format makes merging and comparing datasets easier.

According to Wickham, there are five key ways that data can be "messy." Luckily, the "tidyr" package provides functions that are specifically designed to handle all of these situations. We will focus specifically on three out of the five issues. The other two are more about types of observational units and dataset structures, which are outside the scope of our current analysis. However, I recommend reading the original paper by Wickham for a comprehensive understanding of all the types of data tidiness issues (Wickham 2014). The three issues we will tackle with "tidyr" are the following:

1. Column headers are values, not variable names.
2. Multiple variables are stored in one column.
3. Variables are stored in both rows and columns.

3.7.1 *Loading in the Gapminder Dataset*

So far, our research has focused solely on data about the films themselves. However, we may want to go beyond just analyzing the films themselves and explore the larger political and economic factors that influence the American film industry. For instance, we may be interested in examining the relationship between the industry's growth and the US overall development as a nation. For this, we can use the Gapminder dataset, which is a remarkable collection of global development statistics that was popularized by the Swedish physician, academic, and public speaker Hans Rosling. This comprehensive dataset encompasses a wide array of indicators, such as life expectancy, gross domestic product (GDP) per capita, and population figures for various countries spanning several decades.

We will use a wide version of the Gapminder dataset put together by the Carpentries organization, which will allow us to look at how many real-world datasets are structured and how we can reshape them using tools from the "tidyverse"

(Zimmerman et al., n.d.). The dataset itself can be loaded using the `read_csv()` function, just as we have been doing with our other datasets:

```
# Load the gapminder dataset
gapminder_data <- read_csv("data/ch2_gapminder.csv")
```

As you can see, we have 142 rows (one for each country) and 38 different columns. The first two columns represent the following values:

country—The name of the country
continent—The name representing the continent of the country
The following 36 columns represent three different variables for each of the 12 5-year intervals between 1952 and 2007:
gdpPercap_Y—The GDP per capita for year Y (e.g., "gdpPercap_1952," "gdpPercap_1957"), which represents the size of the economy relative to the population
lifeExp_Y—The life expectancy of the people in the country for the year Y (e.g., "lifeExp_1952," "lifeExp_1957"), which is an indicator of the average living conditions
pop_Y—The total population of the country in the year Y (e.g., "pop_1952," "pop_1957").

This data will need to be transformed into a "tidy" format to match our previous datasets and make it comparable. We'll need to reshape the data to have a format where every row is an observation and each observation has its own dedicated column.

3.7.2 Fixing the Issue of Column Headers as Values

To analyze the relationship between films and economic development in America, we need to connect our film dataset with the Gapminder dataset. The Gapminder dataset, however, is in a wide format, with each column representing a different year and the corresponding development indicators. This is in contrast to our film dataset, which is in a long format, with each row representing a single film and its attributes. These two terms often overlap, but it is important to note that tidy data and long data are not the same thing, although they often overlap. Tidy data is focused on representing the relationships between the values, while long data is focused on the display of those relationships. In fact, we will see later where a "wide" format dataset is tidy.

The Gapminder dataset we are currently examining presents a common challenge in data analysis: a single conceptual variable—gross domestic product (GDP) per capita—is distributed across multiple columns, each representing a different year. In its present wide format, each year's GDP per capita has its own column, such as "gdpPercap_1952" and "gdpPercap_1957." To tidy this data, we need to gather all of the year columns into a single variable and a separate variable for the GDP per capita values. We can achieve this using the `pivot_longer()` function from the "tidyr" package. It transforms wide data into long data by consolidating multiple columns into two columns: one for the "key" (year in our case) and the other for the "value" (GDP per capita).

To apply the `pivot_longer()` function, we use the "cols" argument to indicate which columns in the dataset should be gathered into the "key" column. In our case, we need to gather all the columns that represent different years of GDP per capita data. Instead of listing every column individually, we can use the helper function `starts_with("gdp")`, which automatically selects all columns whose names begin with the string "gdp." This is an efficient way to target a set of related columns without manually typing each one.

The "names_to" argument specifies the name for the new column that will contain the consolidated "key" values. This argument helps us to effectively label the dimension that represents time within our dataset. Before invoking the `pivot_longer()` function, we will also use the `select()` function to isolate only the necessary columns from the original dataset using the `starts_with()` function. By doing so, we ensure that unrelated data is not included in the transformation, which streamlines the process and keeps the focus on the variables of interest:

```
# Create a tidy version of the gapminder dataset
gapminder_tidy <- gapminder_data |>
  select(country, continent, starts_with("gdp")) |>
  pivot_longer(
    cols = starts_with("gdp"),
    names_to = "year",
    values_to = "amount"
  )

# View the structure of the tidy dataset
glimpse(gapminder_tidy)
```

```
Rows: 1,704
Columns: 4
$ country   <chr> "Algeria", "Algeria", "Algeria", "Algeria", "Algeria", "Alge~
$ continent <chr> "Africa", "Africa", "Africa", "Africa", "Africa", "Africa", ~
$ year      <chr> "gdpPercap_1952", "gdpPercap_1957", "gdpPercap_1962", "gdpPe~
$ amount    <dbl> 2449.0082, 3013.9760, 2550.8169, 3246.9918, 4182.6638, 4910.~
```

Upon completing the transformation of our dataset, we've efficiently restructured the data into three succinct columns: "country," "year," and "amount." The "country" column identifies the nation, the "year" column denotes the time, and the "amount" column reflects the per capita GDP values. Nevertheless, we encounter a subtle yet significant issue with the "year" column. Originally, the dataset's column names included both a descriptor of the data type—"gdpPerCap_"—followed by the year it pertains to, such as "gdpPerCap_1952." The `pivot_longer()` operation has consolidated these column names into single entries in the "year" column; however, the entries retain their prefixes, thus conflating the type of data with its temporal identifier.

To isolate the year and ensure that each cell contains a unique and properly formatted value, we can employ the "names_prefix" argument within the `pivot_longer()` function. By specifying the common prefix (`"gdpPerCap_"`) in the "names_prefix" argument, `pivot_longer()` can automatically strip this prefix from the "year" column's entries, leaving us with clean numeric year values as intended. For example, using `names_prefix = "gdpPerCap_"` will transform "gdpPerCap_1952" into "1952" within the "year" column, effectively disentangling the date from the descriptor. Let's update our code to include this refinement:

```
#Adjust the pivot to correctly format the 'year' column and remove the
↪  prefix
gapminder_tidy <- gapminder_data |>
  select(country, continent, starts_with("gdp")) |>
  pivot_longer(
    cols = starts_with("gdp"),
    names_to = "year",
    names_prefix = "gdpPerCap_",
    values_to = "amount"
    )

#View the structure of the updated dataset
glimpse(gapminder_tidy)
```

```
Rows: 1,704
Columns: 4
$ country   <chr> "Algeria", "Algeria", "Algeria", "Algeria", "Algeria", "Alge~
$ continent <chr> "Africa", "Africa", "Africa", "Africa", "Africa", "Africa", ~
$ year      <chr> "gdpPercap_1952", "gdpPercap_1957", "gdpPercap_1962", "gdpPe~
$ amount    <dbl> 2449.0082, 3013.9760, 2550.8169, 3246.9918, 4182.6638, 4910.~
```

The long format of a dataset, which we achieve through the `pivot_longer()` function, is better suited for the type of analyses we want to conduct. However, there are scenarios where one might need to revert to a wide format. This is where the `pivot_wider()` function comes into play. Part of the "tidyr" package, `pivot_wider()` is the counterpart to `pivot_longer()` and is specifically designed to spread a long format dataset into a wide format.

It takes two arguments. The first is a "names_from" argument, which indicates the column in the long format dataset that will provide the names for the new columns in the wide format. For instance, if we have a "year" column that we are transforming back to a wide format, the unique values in "year" will become the names of the new columns. The second argument, "values_from," determines which column in the long format dataset contains the values to be spread across the newly created columns as specified by "names_from." In the context of our Gapminder example, this would be the column containing the actual GDP per capita figures, which will populate the cells of the new wide-format columns.

Additionally, `pivot_wider()` offers the "names_prefix" argument, which can be particularly beneficial when the original wide-format column names included a prefix. In our case, adding "gdpPercap_" before each year in the column names clarifies that the values represent GDP per capita and not another metric:

```
# Convert the 'gapminder_tidy' long format dataset back into a wide
↪   format dataset
gapminder_wide <- gapminder_tidy |>
  pivot_wider(
    names_from = year,
    values_from = amount
  )

# View the structure of the wide format dataset
gapminder_wide
```

```
# A tibble: 142 x 14
   country continent gdpPercap_1952 gdpPercap_1957 gdpPercap_1962 gdpPercap_1967
   <chr>   <chr>              <dbl>          <dbl>          <dbl>          <dbl>
 1 Algeria Africa             2449.          3014.          2551.          3247.
 2 Angola  Africa             3521.          3828.          4269.          5523.
 3 Benin   Africa             1063.           960.           949.          1036.
 4 Botswa~ Africa              851.           918.           984.          1215.
 5 Burkin~ Africa              543.           617.           723.           795.
 6 Burundi Africa              339.           380.           355.           413.
 7 Camero~ Africa             1173.          1313.          1400.          1508.
 8 Centra~ Africa             1071.          1191.          1193.          1136.
 9 Chad    Africa             1179.          1308.          1390.          1197.
10 Comoros Africa             1103.          1211.          1407.          1876.
# i 132 more rows
# i 8 more variables: gdpPercap_1972 <dbl>, gdpPercap_1977 <dbl>,
#   gdpPercap_1982 <dbl>, gdpPercap_1987 <dbl>, gdpPercap_1992 <dbl>,
#   gdpPercap_1997 <dbl>, gdpPercap_2002 <dbl>, gdpPercap_2007 <dbl>
```

In most cases, you won't always have to do all three steps we performed, and the kind of cleanup necessary will depend on how your data was initially recorded and stored. Nonetheless, having these tools at your disposal will make it far easier to tackle the variety of challenges you'll encounter in real-world cultural analytics data.

3.7.3 Fixing Issue with Multiple Variables Stored in One Column

Having successfully applied the `pivot_longer()` function to restructure our GDP data from a wide to a long format, we now turn our attention to the life expectancy and population columns in the Gapminder dataset. Two things to remember: first, rather than typing each column, we can use the `starts_with()` function to specify that we want to gather all columns starting with "lifeExp" and "pop"; and second, we will start with the original data we imported before we began transforming and tidying. Consider the following revised code snippet for tidying the rest of the dataset:

```
# Create a revised tidy version of the gapminder dataset
gapminder_tidy <- gapminder_data |>
  select(country, continent, starts_with("gdp"), starts_with("lifeExp"),
  ↪  starts_with("pop")) |>
  pivot_longer(
    cols = c(starts_with("gdp"), starts_with("lifeExp"),
    ↪  starts_with("pop")),
    names_to = "type",
    values_to = "amount"
  )

# Use glimpse() to get a quick overview of the structure of the
↪  'gapminder_tidy' data frame
glimpse(gapminder_tidy)
```

```
Rows: 5,112
Columns: 4
$ country   <chr> "Algeria", "Algeria", "Algeria", "Algeria", "Algeria", "Alge~
$ continent <chr> "Africa", "Africa", "Africa", "Africa", "Africa", "Africa", ~
$ type      <chr> "gdpPercap_1952", "gdpPercap_1957", "gdpPercap_1962", "gdpPe~
$ amount    <dbl> 2449.008, 3013.976, 2550.817, 3246.992, 4182.664, 4910.417, ~
```

You will notice that there are a couple of problems with this dataset. Unlike before, we can't simply use the "names_prefix" argument since each of the prefixes represents valuable information about the "amount" column. In other words, each entry within this column encodes two distinct pieces of information: the type of measurement (such as GDP per capita, life expectancy, or population) and the corresponding year of the measurement. However, this dual representation violates the tidy data principle that each variable should be contained in its own column. This was not a problem before, since we were only looking at the GDP, so we could safely remove the prefix. The scenario we now have with our data is the second most common issue with "messy data": one column contains multiple variables.

We need to separate this information into two separate columns using the `separate()` function. It contains three arguments. First, the "col" argument designates the column to be separated. In our example, this is the "type" column, where the bundled-up values of measurement type and year need to be divided. Second, the "into" argument identifies the names of the new columns that will result from the separation. It takes a vector of strings as input, with each string representing the name of a new column in the order that the components are expected to be separated. Therefore, we might specify `into = c("measurement", "year")`, which will create two separate columns: one for the type of measurement and one for the year. Finally, the "sep" argument determines the character or pattern that `separate()` will use to identify where the split should occur. Since our combined strings are connected with an underscore (as in "gdpPercap_1952"), we specify `sep`

= "_", indicating that the underscore is the delimiter between the pieces of data we wish to extract:

```
# Separate the 'type' column into two new columns: 'measurement' and
↪ 'year'
gapminder_tidy <- gapminder_tidy |>
  separate(
    col = type,
    into = c("measurement", "year"),
    sep = "_"
  )

# Use the glimpse function to get a quick look at the structure and
↪ contents of the updated dataset
glimpse(gapminder_tidy)
```

```
Rows: 5,112
Columns: 5
$ country     <chr> "Algeria", "Algeria", "Algeria", "Algeria", "Algeria", "Al~
$ continent   <chr> "Africa", "Africa", "Africa", "Africa", "Africa", "Africa"~
$ measurement <chr> "gdpPercap", "gdpPercap", "gdpPercap", "gdpPercap", "gdpPe~
$ year        <chr> "1952", "1957", "1962", "1967", "1972", "1977", "1982", "1~
$ amount      <dbl> 2449.008, 3013.976, 2550.817, 3246.992, 4182.664, 4910.417~
```

We see that by separating the "type" column, we gain the "measurement" and "year" columns, which tidies the data significantly. Now, each row contains a single observation for a given country, for a specific year, and for a particular measurement type (GDP per capita, life expectancy, or population). However, there is still one more issue to address: we have variables stored as both row values and variable names.

3.7.4 Fixing the Issue of Variables Being Stored in Both Rows and Columns

According to Wickham, the "most complicated form of messy data occurs when variables are stored in both rows and columns" (Wickham 2014). We have this in our data as a single observational unit—representing a particular country in a particular year—is distributed across multiple rows. In a tidy dataset, each observation should be contained within a single row to facilitate analysis.

To coalesce these scattered fragments of data into cohesive rows, we must reshape our dataset from its current long format back into a wide format, where each observational unit is represented by a single row, with separate columns for each measurement type using the `pivot_wider()` function we saw earlier:

```
# Transform the 'gapminder_tidy' long format dataset back into a wide
↪  format dataset
gapminder_tidy <- gapminder_tidy |>
  pivot_wider(
    names_from = "measurement",
    values_from = "amount"
  )

# Take a glimpse at the final data
glimpse(gapminder_tidy)
```

```
Rows: 1,704
Columns: 6
$ country   <chr> "Algeria", "Algeria", "Algeria", "Algeria", "Algeria", "Alge~
$ continent <chr> "Africa", "Africa", "Africa", "Africa", "Africa", "Africa", ~
$ year      <chr> "1952", "1957", "1962", "1967", "1972", "1977", "1982", "198~
$ gdpPercap <dbl> 2449.0082, 3013.9760, 2550.8169, 3246.9918, 4182.6638, 4910.~
$ lifeExp   <dbl> 43.077, 45.685, 48.303, 51.407, 54.518, 58.014, 61.368, 65.7~
$ pop       <dbl> 9279525, 10270856, 11000948, 12760499, 14760787, 17152804, 2~
```

After all that we finally have a tidy version of the original Gapminder data. In most cases, you won't always have to do all three steps we performed, and the kind of cleanup necessary will depend on how your data was initially recorded and stored. Nonetheless, having these tools at your disposal will make it far easier to tackle the variety of challenges you'll encounter in real-world cultural analytics data.

3.8 Tidying the Film Dataset

Now that we have learned to tidy the Gapminder dataset, let's apply these concepts to our original dataset. Luckily, we do not have nearly as many problems with this dataset. Each observation is already a row, and each variable is already a column.

Nonetheless, we still have a problem similar to the one that we saw earlier in the Gapminder dataset with the measurement and year columns. Each cell does not necessarily contain only one value since we have a single column containing multiple genres. If we want to analyze how genre affects other attributes in the dataset, it would be helpful to separate these genres. However, simply separating the genres into multiple columns is not easy since not all films are listed with the same amount of genres. How we split this column depends on the research question we hope to explore. Should we just count the first genre listed as the definitive genre for the film? Should a film that has two different genres count for both genres, essentially counting it twice? Or should we create new columns holding each genre so that the first is in a column called genre_1, the second in genre_2, and so on? These are questions that you need to consider when you are faced with similar situations.

For our analysis, we will go with the first option and only count the first listed genre as the one belonging to each film. To do this, we will split our dataset into different rows based on the genre and choose the first row corresponding to each title. To split a table into multiple rows based on a column, we can use the `separate_rows()` function. This function takes two arguments. The first argument is the column that we want to split, which in this case is "genre." The second argument is the character that we want to split by. In this case, we have genres separated by the symbol `,`, so we will set the second argument, the separator, as `','`.

This step will be challenging due to the myriad of operations we need to do. Go slow, take your time, review, and break down the problem into smaller, manageable tasks. We will begin by using the `separate_rows()` function to create multiple rows with one genre each. This means that if a movie has three genres listed, it will show up three times: once for each genre. Since we will then have only one genre for each column, we will rename the "genres" column to "genre" to reflect the change. We will also remove any films that don't have a genre using `drop_na()`.

We will then use the `group_by()` function to let R know that the rest of our operation should be done for each group rather than the dataset as a whole. We will then use the `slice()` function to choose only a select amount of rows. The "n" argument to the `slice()` function tells us how many rows we want to keep. By passing one, we ensure that we only keep the first row for each film, effectively selecting only the first listed genre. If you are comfortable with counting the films multiple times, you could simply remove the `slice()` call. Finally, we will use the `ungroup()` function so that future operations apply to the entire dataset and not just the grouped subsets:

```
# Separate the 'genres' column into multiple rows, keeping only the first
↪  genre for each film. Rename the column to 'genre' and remove any rows
↪  with missing values.
films_tidy <- films |>
  separate_rows(genres, sep = ",") |>
  rename(genre = genres) |>
  drop_na() |>
  group_by(title) |>
  slice(1) |>
  ungroup()

# Glimpse the tidied film dataset

glimpse(films_tidy)
```

```
Rows: 15,424
Columns: 10
$ title                <chr> "#1 Cheerleader Camp", "#Captured", "#Horror"~
$ tmdbId               <dbl> 41371, 476424, 301325, 632076, 359732, 419627~
$ rating               <dbl> 2.000000, 2.250000, 2.105263, 2.500000, 1.750~
$ runtime              <dbl> 95, 81, 90, 93, 80, 85, 90, 84, 83, 99, 98, 9~
$ release_year         <dbl> 2010, 2017, 2015, 2019, 2015, 2016, 2018, 201~
```

```
$ decade                    <chr> "2010s", "2010s", "2010s", "2010s", "2010s", ~
$ decade_average_runtime    <dbl> 93.60059, 93.60059, 93.60059, 93.60059, 93.60~
$ decade_runtime_deviation <dbl> 1.3994056, 12.6005944, 3.6005944, 0.6005944, ~
$ runtime_hours             <dbl> 1.583333, 1.350000, 1.500000, 1.550000, 1.333~
$ genre                     <chr> "Comedy", "Horror", "Drama", "Thriller", "Com~
```

3.9 Exploring the Relationship Between Films and Economic Development

We now have a fairly substantial dataset that allows us to ask interesting questions. Let's go ahead and see how we can use what we have learned to understand our data better. Normally, when doing exploratory data analysis, it is beneficial to use visualizations, and we will see how the "ggplot2" package allows us to do this within the same tidy framework in the next chapter. For now, however, we will simply print out the results of our analyses.

3.9.1 Most Popular Genre

Since we tidied the dataset regarding genre, we can quickly look at the most popular genre. Remember that we are only looking at the first genre listed for each film since that is the method we chose earlier. We will group our data by genre and then count how many films belong to each. After that, we can sort the data in descending order to determine the most popular genre.

We can use the `count()` function to make this task even more accessible, as it automatically counts the number of observations in each group. This function is a shorthand for a call to `group_by()` and `n()`. The counts will be in the "n" column to demonstrate this:

```
# Count the number of films in each genre and sort the results in
↪  descending order
genre_counts <- films_tidy |>
  count(genre, name = "number_of_films") |>
  arrange(desc(number_of_films))

# Print the genre counts
genre_counts
```

```
# A tibble: 19 x 2
   genre            number_of_films
   <chr>                      <int>
 1 Comedy                      3291
 2 Drama                       3110
 3 Documentary                 1978
```

```
 4 Horror                    1529
 5 Action                    1337
 6 Thriller                   785
 7 Crime                      483
 8 Romance                    462
 9 Family                     403
10 Adventure                  373
11 Science Fiction            344
12 TV Movie                   297
13 Animation                  281
14 Music                      202
15 Fantasy                    199
16 Mystery                    169
17 Western                     79
18 War                         52
19 History                     50
```

The analysis of our dataset reveals that Comedy emerges as the most prevalent genre, boasting a total of 1205 films. This genre's popularity is significantly higher compared to others in the dataset. Following closely behind is Drama, with a substantial count of 1068 films, showcasing its strong presence and appeal among audiences. Horror, while not as dominant, still displays a considerable representation with 599 films. Why is this the case? What cultural, social, or economic factors contribute to the prevalence of these genres in American cinema? These questions can serve as the foundation for further exploration and analysis.

3.9.2 *Genres and Ratings*

To analyze how genres are related to ratings, we can use the "tidyverse" tools to group our data by genre and then calculate the mean rating for each genre. This will allow us to understand if the number of films in a genre is related to the average rating of that genre. In other words, are filmmakers producing more films in genres that they believe are more popular with audiences?

```
# Group the data by genre and calculate the mean rating for each genre.
↪  Arrange the results in descending order.
genre_ratings <- films_tidy |>
  group_by(genre) |>
  summarize(mean_rating = mean(rating, na.rm = TRUE)) |>
  arrange(desc(mean_rating))

# Print the genre ratings
genre_ratings
```

```
# A tibble: 19 x 2
   genre           mean_rating
   <chr>                 <dbl>
 1 Documentary            3.39
 2 Music                  3.32
 3 History                3.18
 4 Drama                  3.07
 5 Romance                2.99
 6 TV Movie               2.98
 7 Crime                  2.96
 8 Comedy                 2.93
 9 Animation              2.92
10 Adventure              2.92
11 War                    2.89
12 Western                2.86
13 Mystery                2.79
14 Fantasy                2.77
15 Family                 2.75
16 Thriller               2.73
17 Action                 2.66
18 Science Fiction        2.55
19 Horror                 2.34
```

Our dataset analysis indicates that Documentary films stand out with the highest average rating of 3.43, suggesting a strong appreciation for this genre among viewers. Close behind are History films, which maintain a notable average rating of 3.4. This proximity in ratings points to a shared value placed on educational and informative content by audiences. On the opposite end of the spectrum, Horror films record the lowest average rating at 2.47, highlighting a distinct disparity in audience reception.

This intriguing trend prompts further examination into the elements that contribute to these ratings. For instance, what specific characteristics of Documentary and History films resonate so well with audiences, driving up their average ratings? Is it the authenticity, educational value, or perhaps the narrative style that captivates viewers? Conversely, identifying the reasons behind the lower ratings of Horror films can be equally enlightening. Are these films failing to meet viewer expectations in terms of suspense, plot development, or production quality?

However, one reason may be that there are fewer Documentary and History films in the dataset compared to other genres. This could mean that the ratings for these genres are more volatile since they are based on fewer films. To understand this better, we can merge our "genre_ratings" dataset with the "genre_counts" dataset we created earlier. This will allow us to compare the number of films in each genre with the average rating for that genre:

```
# Merge the genre ratings and genre counts datasets
genre_analysis <- genre_ratings |>
  left_join(genre_counts, by = "genre")

# Print the merged dataset
genre_analysis
```

```
# A tibble: 19 x 3
   genre           mean_rating number_of_films
   <chr>                 <dbl>           <int>
 1 Documentary            3.39            1978
 2 Music                  3.32             202
 3 History                3.18              50
 4 Drama                  3.07            3110
 5 Romance                2.99             462
 6 TV Movie               2.98             297
 7 Crime                  2.96             483
 8 Comedy                 2.93            3291
 9 Animation              2.92             281
10 Adventure              2.92             373
11 War                    2.89              52
12 Western                2.86              79
13 Mystery                2.79             169
14 Fantasy                2.77             199
15 Family                 2.75             403
16 Thriller               2.73             785
17 Action                 2.66            1337
18 Science Fiction        2.55             344
19 Horror                 2.34            1529
```

The merged dataset provides a comprehensive overview of the relationship between the number of films in each genre and the average rating for that genre. By juxtaposing these two metrics, we can discern patterns and trends that illuminate the dynamics of the American film industry. For instance, the stark contrast between the high average rating of History and the relatively low number of films in this genre underscores the genre's niche appeal and the discerning tastes of its audience. Conversely, the widespread popularity of Comedy films, coupled with their moderate average rating, suggests a broad audience base that appreciates the genre's diverse offerings.

It's worth noting that these results are specific to the dataset we are working with and may not necessarily reflect the preferences of all film viewers. Additionally, there are choices made about which genres should be listed as separate. For instance, some may see "Crime" movies as a subgenre of "Drama." If we had not restricted genre to a single genre per film, we might see overlaps influencing the ratings.

3.9.3 Did the Film Industry Grow with the American Economy?

To delve into the correlation between the film industry and America's economic advancement, we can align the number of films produced each year with the nation's per capita GDP. However, our task is met with a constraint—the Gapminder dataset at our disposal reports the GDP per capita at 5-year intervals and only up to the year 2007. To accommodate this, we need to curate our film data into a compatible format that mirrors this 5-year cadence.

To do this, we will filter it to focus exclusively on films released during the period from 1962 to 2007, ensuring synchronicity with the time span for which we have Gapminder data. Next, we will categorize the films based on their release into distinct 5-year brackets. To achieve this, we will introduce a new attribute to our dataset, which we'll refer to as the "five_year_interval." To do this, we can use the `%/%` operator. If you remember from Chap. 1, this operator allows us to perform integer division. In other words, it divides the numbers and returns the quotient without the remainder. We will assign each film to its 5-year segment by subtracting 1962 from the film's release year, doing integer division by 5, and rescaling.

Once our films are neatly grouped, we proceed to quantify them. Using the `count()` function, we will tally up the films within each interval, a process that culminates in the creation of a new column aptly named "films_released." This new column serves as a quantifiable metric of the industry's output over the designated intervals:

```
# Filter the film data to focus on films released between 1962 and 2012
films_five_years <- films_tidy |>
  filter(release_year >= 1962 & release_year <= 2007) |>
  mutate(
    year = ((release_year - 1962) %/% 5 * 5) + 1962
  ) |>
  count(year, name = "films_released")

# Print the films released in five-year intervals
films_five_years
```

```
# A tibble: 5 x 2
   year films_released
  <dbl>          <int>
1  1987            533
2  1992           1497
3  1997           1662
4  2002           1887
5  2007            529
```

We can now fuse our film data with the economic data housed within the Gapminder dataset. To meld these two repositories of information, we will use the `left_join()` function we saw earlier, deftly matching the years from our 5-year interval column with the corresponding years in the Gapminder dataset. To make it easier to read, we will also filter the Gapminder data to only include the United States. Because the Gapminder data has "year" as a character, we need to convert it to a numeric value to match the "year" in our film data:

```
# Filter the Gapminder data to focus on the United States
gapminder_us <- gapminder_tidy |>
  filter(country == "United States") |>
  mutate(year=as.numeric(year))
```

```
# Merge the film data with the Gapminder data
film_economic_data <- films_five_years |>
  left_join(gapminder_us, by = "year")

# Print the merged dataset
film_economic_data
```

```
# A tibble: 5 x 7
   year films_released country       continent gdpPercap lifeExp       pop
  <dbl>          <int> <chr>         <chr>         <dbl>   <dbl>     <dbl>
1  1987            533 United States Americas     29884.    75.0 242803533
2  1992           1497 United States Americas     32004.    76.1 256894189
3  1997           1662 United States Americas     35767.    76.8 272911760
4  2002           1887 United States Americas     39097.    77.3 287675526
5  2007            529 United States Americas     42952.    78.2 301139947
```

After examining the data, there is a noticeable upward trend in the number of films produced that corresponds with rises across several of America's key socioeconomic indicators: GDP per capita, population, and life expectancy. This parallel growth suggests that the film industry is potentially intertwined with the nation's broader economic expansion. As the nation's economy grows, it's plausible that increased prosperity furnishes the means for more film production through greater investments, a more robust infrastructure for filmmaking, and a larger consumer base able to financially support the industry. Additionally, a growing and healthier population might contribute to a larger audience for entertainment.

The upward trajectory in life expectancy could also reflect improvements in social conditions and public health that, in turn, provide a more stable environment for the arts and entertainment sectors to flourish. Nonetheless, while the observed correlations are promising, they do not suffice to establish causality or a definitive linkage. To move beyond a mere surface understanding and to ascertain the precise nature of the relationship between the film industry's growth and America's economic development, a more in-depth analysis is warranted. It would be prudent to consider additional variables such as government policies, international trade, technological advances, and cultural trends that might affect both the economy at large and the film industry specifically.

3.10 Conclusion

This chapter has provided a glimpse into cleaning, analyzing, and exploring data using "tidyverse" tools such as "dplyr" and "tidyr." We have demonstrated various techniques to manipulate and analyze our dataset, including information on American films. Our exploration has revealed valuable insights into the industry trends, popular genres, and the relationship between IMDb ratings and genres. Moreover, we identified a potential correlation between the growth of the American film industry and the US economic growth. Further research and analysis can be conducted to dive deeper into the factors contributing to the industry's growth and the intricacies of the relationship between film success and audience preferences.

Leveraging "tidyverse" tools and data manipulation techniques can take some practice, but with time and experience, they become invaluable assets in data analysis. By applying these techniques and tools in your data analysis projects, you can efficiently extract valuable insights and make data-driven decisions. Remember that every dataset is different, and the methods used in this chapter may require adjustments depending on your analysis's specific goals and requirements. The key lies in understanding the underlying logic and concepts and adapting them to suit your needs.

In the following chapters, we will see how the "tidyverse" allows us to visualize our data using "ggplot2," an advanced and flexible plotting library. In particular, we will learn how to create a variety of plots and customize them to extract insights, explore patterns and trends, and communicate our findings effectively.

Chapter 4
Understanding the Grammar of Graphics

Abstract Delving into the grammar of graphics, this chapter unravels the complexities of data visualization in cultural analytics, emphasizing its interpretative nature. The exposition is grounded in the theoretical underpinnings of "ggplot2," inspired by Leland Wilkinson's pioneering framework, illustrating the construction of graphical elements through logical principles that mirror grammatical constructs in language. Key elements of the grammar of graphics are explored—from the manipulation of data and aesthetics to the strategic use of geometric objects, statistics, and themes—equipping researchers with the acumen to translate complex cultural data into persuasive visual stories. The section culminates with the application of "ggplot2" to the Musical Sentiment Dataset (MuSe), using practical examples and exercises to showcase the tool's capacity for dissecting the emotional impact of music across genres.

Keywords Grammar of graphics · ggplot2 · Data visualization · Interpretative visualization · Statistical storytelling · Visual narratives

4.1 Introduction

In cultural analytics, researchers are tasked with a formidable challenge: to make sense of the intricate patterns that define human creativity and cultural evolution. This pursuit requires a keen eye for detail and an ability to synthesize complex information into a comprehensible format that can be shared and understood to broader audiences. Consequently, visualizing data is not merely a technical step. It is an interpretative act that illuminates the hidden structures of social norms, revealing relationships and dynamics that may not be immediately apparent through raw data alone.

In Chap. 2, we explored how base R facilitates data visualization with its robust `plot()` function. This function is pivotal in the early stages of data exploration and will continue to be a valuable asset for your research. However, as you advance in your analytical journey, you will likely observe a significant trend: most R users gravitate toward "ggplot2" (Wickham 2016). "ggplot2" has a plethora of

N. Siddiqui, *Cultural Analytics in R: A Tidy Approach*, Use R!,
https://doi.org/10.1007/978-3-031-96618-7_4

available extensions that enrich its functionality and adapt it for domain-specific applications that are not readily accommodated by other software languages. This wide-ranging support has effectively elevated "ggplot2" from a mere plotting tool to an indispensable instrument for data storytelling in cultural analytics.

The power and advantage of "ggplot2" come from its theoretical basis in Leland Wilkinson's seminal work, *The Grammar of Graphics* (Wilkinson and Wills 2005). In his work, Wilkinson posits that graphical representations can be built from logical principles, just as language is constructed from an interplay of grammatical rules. All too often, R books and tutorials emphasize the "how-to" aspect of creating charts and graphs to showcase the package's rich capabilities with ready-to-use code snippets. While this can provide immediate satisfaction and impressive results, it glosses over the "why" and "when"—understanding underlying principles that inform effective visualization choices. This leads to later frustration as users encounter more complex data scenarios or bespoke visualization needs that are not quickly addressed with a superficial understanding.

To mitigate this issue, this chapter takes a pedagogical detour through the foundational concepts that fuel "ggplot2's" engine. By grounding ourselves in its theoretical framework, we establish a platform from which we can dive into R's implementation of the grammar of graphics. Thus, when the time comes to craft complex and nuanced visualizations, you will possess not just the technical know-how but also the theoretical insight to tackle the challenge with confidence and creativity. After overviewing the grammar of graphics, we will do a detailed examination of each "layer" in it by paying particular attention to how layers interact and build upon each other. We'll see how this allows for a high degree of customization, which is critical in a field as nuanced as cultural analytics. Through practical examples and guided exercises, you will learn how "ggplot2" complements the tidy approach to data analysis.

4.2 The Rise of the Grammar of Graphics

To understand R's implementation of the "grammar of graphics," it is important to have a general foundation of how a theory of graphs emerged. During the nineteenth century, statisticians attempted to develop a standardized framework to understand how we process visual information (Palsky 1999). For these researchers, visualization served as a more efficient tool for explanation than numbers, and they saw it as one that could cross national boundaries. However, during the latter half of the nineteenth century, the International Statistical Institute stressed that it was ultimately better to focus on standardizing data rather than the visualizations themselves by arguing that the validity of the latter merely stemmed from the strength of the former (Brown 1872).

The viewpoint that good visualization was ultimately a byproduct of data itself held sway throughout much of the twentieth century. However, in 1967, Jacques Bertin challenged this notion with the release of his *Semiology of Graphics*, a critical text that resurrected attempts to standardize graphic representation (Bertin 1983). Bertin stressed the separation of content from visualization and created a typology for different parts of a graph. He saw a graph as being divided into two key elements: planar variables and retinal variables. The planar variables focused on coordinates of the graph, while the retinal variables were properties such as shape and size of symbols used to represent data. For our purposes, the exact properties of the graph Bertin chose are less relevant than his overall idea that a general framework for graphics could exist.

Bertin's implementation proved increasingly challenging with the rise of statistical computing. For him, the ideal graph was a sheet of paper with the statistician ultimately responsible for doing any calculations. In response, many researchers sought to create new standards more attuned to the contemporary computational environment. Although some of these became popular in domain-specific academic research, it was not until Leland Wilkinson's 1999 release of *The Grammar of Graphics* that the computational statistics community began to coalesce around a single framework (Wilkinson and Wills 2005).

Wilkinson's interest in graphics emerged from his work in psychiatric research. After receiving his PhD in psychology from Yale in 1975, he began working at the University of Illinois at Chicago as an assistant professor of psychology. There, he wrote a statistical software package entitled SYSTAT. As he notes, he was "determined to produce a package that could draw every statistical graphic [he] had ever seen" (Wilkinson and Wills 2005, ix). Researchers lauded the range of graphics the software could produce, and Wilkinson eventually sold the project to SPSS in 1995.

In *The Grammar of Graphics*, Wilkinson expanded much of the theoretical basis that formed SYSTAT. He postulated what he called the Graphics Production Language (GPL), which consisted of key layers that could be manipulated and put together to create a myriad of graphs. Drawing on linguistic theory, he wanted to understand how a graph could be semantically and syntactically correct. Notably, this had little to do with the graph's aesthetic qualities. As he notes, depending on taste, the graphs his system could create could be visually unappealing but they had the benefit of always maintaining meaning.

Wilkinson's theory proved particularly influential to Hadley Wickham, the founder of the "tidyverse." Wickham, recognizing the transformative power of Wilkinson's *Grammar of Graphics*, was inspired to operationalize these theoretical constructs within the R programming environment. He viewed Wilkinson's book as the cornerstone of graphical representation, hailing it as "the most important modern work in graphical grammar" (Wickham 2010a). Driven by a desire to make data visualization both accessible and sophisticated, Wickham channeled Wilkinson's principles into the development of "ggplot2."

While highly inspired by Wilkinson, "ggplot2" differs somewhat from his original formulation. Summarizing the changes, Wickham notes, "The grammar differs from Wilkinson's in its arrangement of the components, the development of a hierarchy of defaults, and in that it is embedded inside another programming language" (Wickham 2010b). The fact that Wickham was implementing the grammar in R meant he could rely on various built-in functions of the language for calculation that Wilkinson could not. Ultimately, this allowed him to simplify the overall grammar.

With small tweaks from contributors online, the overall layers of "ggplot2" have now broadly coalesced around the following layers (although you may find other variations):

1. Data
2. Aesthetics
3. Geometries
4. Facets
5. Statistics
6. Coordinates
7. Themes

Throughout this chapter, we will go through each layer and see how it allows us to create almost all meaningful plot types. To new users, the layered complexity might initially appear daunting. Yet, as we progress, it will become evident that this intricacy is not a burden but rather a powerful ally. "ggplot2" is intuitively designed to operate with a level of autonomy that selectively invokes intelligent defaults whenever a layer's specific details are omitted. This intelligent automation is akin to an experienced co-pilot, stepping in when needed, allowing the analyst to retain command while alleviating the minutiae of decision-making. The result is a harmonious balance: the flexibility to tailor your visual narrative is preserved, unencumbered by the potential verbosity of code.

4.3 Loading the Musical Sentiment Dataset

The data we will work with in this chapter comes from the Musical Sentiment Dataset (MuSe) put together by Christopher Akiki and Manuel Burghardt. It contains information for approximately 90,000 songs that they scraped from the site last.fm along with measures of how they impact emotion. On last.fm, users can tag each song with a list of words. After scraping a sample set of songs, the authors of the MuSe dataset kept only those with at least one mood-related tag. Finally, they mapped the tags to valence, arousal, and dominance while also adding the Spotify ID and MusicBrainz ID. For more information on how these values are calculated, I recommend you check out the excellent write-up by the authors in *The Journal of Open Humanities Data* (Akiki and Burghardt 2021).

We are going to use "ggplot2" to explore the intricate psychological constructs of valence, arousal, and dominance as they harmonize with the diverse genres of music. We will delve into probing questions such as: How do different genres convey emotion through music? Does the intensity of a song correlate with its perceived control or dominance? And what role does culture play in shaping the emotional expression within musical compositions? By employing "ggplot2's" multifaceted graphical system, we will begin to uncover the answers to these questions.

We start by importing our data using the `read_csv()` function and taking a look at it using `glimpse()`:

```
# Import MuSe dataset
muse <- read_csv("data/ch3_muse.csv")

# Preview data structure
glimpse(muse)
```

```
Rows: 90,001
Columns: 11
$ lastfm_url                <chr> "https://www.last.fm/music/eminem/_/%2527till%2~
$ track                     <chr> "'Till I Collapse", "St. Anger", "Speedin'", "B~
$ artist                    <chr> "Eminem", "Metallica", "Rick Ross", "M.I.A.", "~
$ seeds                     <chr> "['aggressive']", "['aggressive']", "['aggressi~
$ number_of_emotion_tags <dbl> 6, 8, 1, 13, 7, 9, 1, 10, 20, 4, 9, 4, 8, 3, 7,~
$ valence_tags              <dbl> 4.550000, 3.710000, 3.080000, 6.555071, 3.77117~
$ arousal_tags              <dbl> 5.273125, 5.833000, 5.870000, 5.537214, 5.34823~
$ dominance_tags            <dbl> 5.690625, 5.427250, 5.490000, 5.691357, 5.44176~
$ mbid                      <chr> "cab93def-26c5-4fb0-bedd-26ec4c1619e1", "727a25~
$ spotify_id                <chr> "4xkOaSrkexMciUUogZKVTS", "3fOc9x06lKJBhz435mIn~
$ genre                     <chr> "rap", "metal", "rap", "hip-hop", "metal", "met~
```

As you can see, each row of the dataset consists of 11 different fields:

- **lastfm_url**: The URL of the song on last.fm.
- **track**: The name of the song.
- **artist**: The artist's name for the song chosen.
- **seeds**: The original songs were determined based on a seed of 279 mood labels from AllMusic. This column contains the original seeds for each song.
- **number_of_emotion_tags**: The number of tags from last.fm that were used to calculate the valence, arousal, and dominance scores.
- **valence_tags**: A measure of pleasantness for the song.
- **arousal_tags**: A measure of intensity for the song.
- **dominance_tags**: A measure of control for the song.
- **mbid**: The identification number for the song on MusicBrainz.
- **spotify_id**: The identification number for the song on Spotify.
- **genre**: The genre of the song.

The diversity of genres, from classical to hip-hop, provides a rich canvas to explore the full spectrum of human emotions in music. With these measures, we can now explore how the tags given by users translate into a deeper understanding of a song's emotional resonance.

4.4 Understanding the Three Major Elements of Grammar

One of the distinct advantages of R's implementation of the grammar of graphics, particularly through "ggplot2," is its propensity for selecting sensible defaults that streamline the creation of visualizations. As such, while there exists a plethora of customizable elements within each graph that we can fine-tune to our aesthetic and analytical desires, the necessity for such adjustments is often minimal. In practice, three fundamental layers form the core infrastructure required to materialize the majority of the graphs you will encounter in statistical storytelling:

1. Data
2. Aesthetics
3. Geometric objects (geoms)

4.4.1 Data

The first part of the grammar of graphics is "data." As the name implies, data holds the data frame or tibble we would like to explore. This is passed to the first argument in the `ggplot()` function, which begins the creation of our graph. Because "ggplot2" is a core package of the "tidyverse," it will load automatically when you load the "tidyverse." Thus, you do not need to load additional packages if you have already loaded the "tidyverse."

Let's look at how the `ggplot()` function works by passing our MuSe dataset to the first layer:

```
# Start basic plot
ggplot(data = muse)
```

Oh no! If you run the above command, you will find that you don't see anything. If you are used to base R's plotting, you may have expected "ggplot2" to provide the beginnings of a visualization. But the function will not display anything until we explicitly state which values to map from our tibble onto different parts of our graph.

Before we start mapping variables to our graph though, it may be helpful to filter out some of our data. Ninety-thousand songs is a lot to analyze, so let's filter by genres we may be interested in. Luckily, "ggplot2" can take data directly from the pipe operator. We will filter for songs that are rap, metal, and funk and run the `ggplot()` function on the corresponding result:

```
#| label: fig-filtered-plot
#| fig-cap: "Emply plot of the MuSe dataset filtered by genre
# Choose three genres and start ggplot
muse |>
```

```
  filter(genre == "rap" | genre == "funk" | genre == "metal") |>
  ggplot()
```

Like before, you will find that we don't see anything as "ggplot2" requires that we explicitly map our variables that we need to display. However, you will find that manipulating data before passing it through the pipe operator to `ggplot()` is fairly common, so it's good to see this early on. As a result, I will use this convention in the rest of the book, so be sure you understand how the data is being sent to the `ggplot()` function before continuing.

4.4.2 Aesthetics

Humanities scholars and social scientists often engage with aesthetics through a distinctly visual lens, focusing on the elements that contribute to the overall appeal and impact of imagery and design. This visual orientation naturally leads to an expectation that "aesthetics" in "ggplot2" should pertain to the decorative aspects of a graph, such as the color palette, the shape of data points, or the style of lines on a scatter plot. Indeed, these are critical elements that influence the visual reception of a graph. However, it is essential to grasp that in "ggplot2," aesthetics serve a more utilitarian purpose. They are the channels through which data variables are assigned—or mapped—to visual properties within the graph. Put more simply, it is where you denote which columns you are interested in plotting.

Before you begin making a graph, think about the data you want to explore. For instance, we may suspect that people tend to find intense songs more pleasurable. One way to check this is by creating a scatter plot with "valence_tags" on the x-axis and "arousal_tags" on the y-axis. Using the `aes()` function, we can make these values explicit in our call to `ggplot()`:

```
# Plot valence vs arousal in filtered dataset
muse |>
  filter(genre == "rap" | genre == "funk" | genre == "metal") |>
  ggplot(mapping = aes(x = valence_tags, y = arousal_tags))
```

You will see that we finally have the beginnings of our graph (Fig. 4.1). We find labels have been created for us that show what tags correspond to the x- and y-axes, respectively. In addition, grid lines with sensible ranges based on our data are also present. However, we still do not have any of our data points. To see them, we need to determine the kind of object we want to use to display each data point.

4.4.3 Geoms

So far, we have created the basic building blocks for our graph in terms of labels and grids. However, we still need to determine the actual objects that we will use

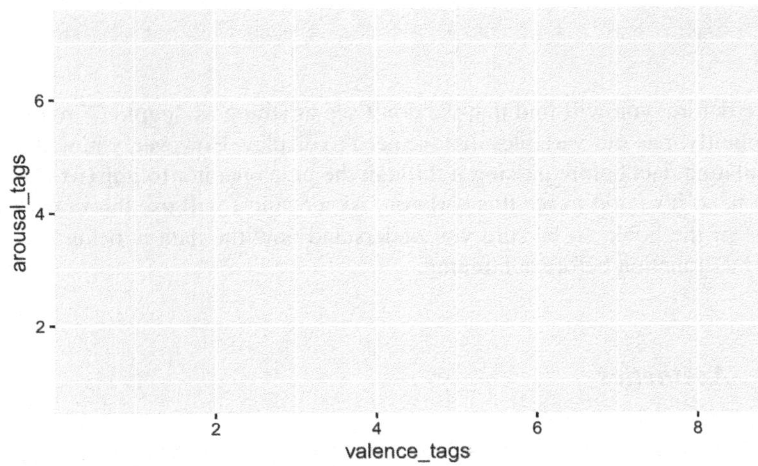

Fig. 4.1 Beginnings of a scatter plot showing the relationship between valence and arousal tags

to represent our data. In "ggplot2," we denote these explicitly through "geoms" or geometric objects.

What is important to remember is that these objects are arbitrary and interchangeable. In a traditional scatter plot, we usually represent data as points, but they could just as well be images, triangles, or squares. The choice of what geometric objects should represent your data is a matter of convention and not necessarily indicative of the theoretical validity of the graph. As a cultural analyst, you should make sure that you are basing your choices on your audience. Thus, while we could use a smiley face emoji or heart shape, most people are probably expecting dots since they are easy to read and understand, so that is what we will use.

To add a new layer to our graph, we will use the + operator. You may wonder why we are not using the pipe operator since that is such a fundamental feature of the "tidyverse." Much of the reason behind this is a historical quirk. The pipe operator works with functions, and the first argument needs to be the object being passed through, which is not the case for all layers. In fact, "ggplot2" actually predates the "tidyverse" and was added to it retroactively. The author of "ggplot2" has stated that had he discovered the pipe operator earlier, he would not have created "ggplot2" since the original ggplot worked in a way that aligned with the pipe operator ("Why Can't Ggplot2 Use." 2018). While some have called for adding the pipe operator into "ggplot2," the trade off in terms of labor and code fragmentation was deemed too great for the purposes of stylistic consistency. In short, when using "ggplot2," you should use the + operator when defining your graph rather than the pipe operator. However, you will find that it is easy to accidentally switch these around. So, if your visualization is not working, check to make sure you used the correct operator.

In most cases, geoms correspond with the "type" of graph you want to create. For instance, to create a histogram, we add `geom_histogram()`. Likewise, to create a

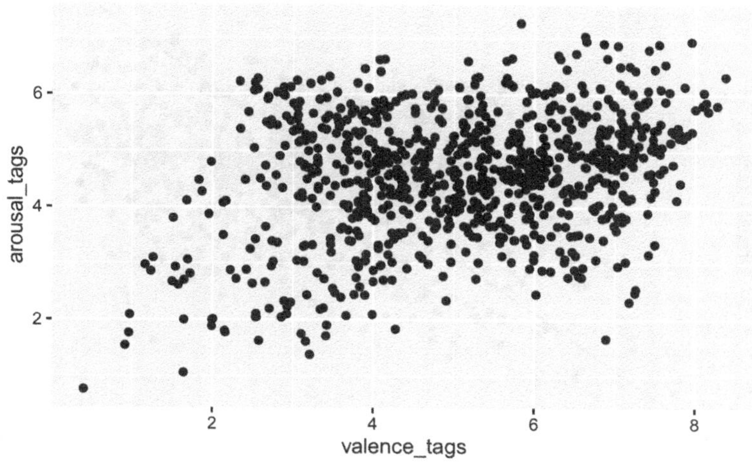

Fig. 4.2 Scatter plot showing the relationship between valence and arousal tags for selected music genres (rap, funk, metal)

box plot, we add `geom_boxplot()`—we will look at each of these in more detail below. Since scatter plots are usually made of points, the layer for creating a scatter plot is `geom_point()`:

```
# Add points for scatter plot to valence vs arousal filtered dataset
muse |>
  filter(genre == "rap" | genre == "funk" | genre == "metal") |>
  ggplot(mapping = aes(x = valence_tags, y = arousal_tags)) +
  geom_point()
```

Figure 4.2 shows the distribution of pleasantness (valence) and intensity (arousal) values. One way to add a reference line to our plot is by using `geom-abline()`, which by default adds a diagonal line with a slope of 1 and intercept of 0. While this isn't a fitted regression line (which we'll explore in our chapter on regression analysis), it can serve as a useful reference point for examining the relationship between these variables:

```
# Create scatter plot with reference line
muse |>
  filter(genre == "rap" | genre == "funk" | genre == "metal") |>
  ggplot(mapping = aes(x = valence_tags, y = arousal_tags)) +
  geom_point() +
  geom_abline()
```

As seen in Fig. 4.3, `geom-abline()` creates a default line with a slope of 1 and intercept of 0 (a 45-degree line). While this demonstrates how to add a line layer to our plot, for a true regression line that fits our data, we would either need to specify

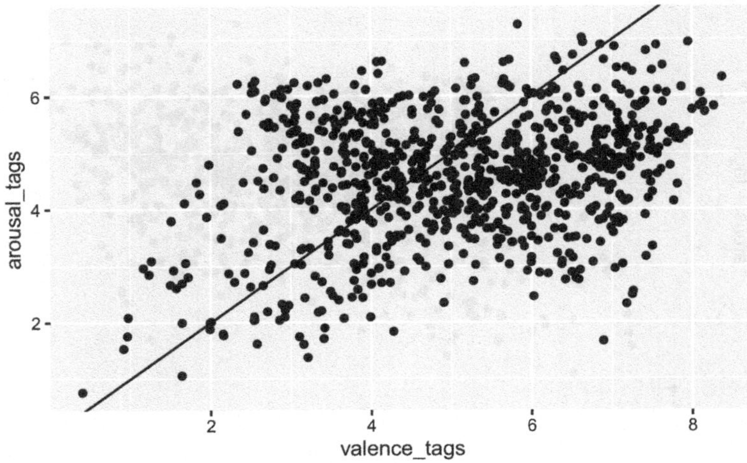

Fig. 4.3 Scatter plot of valence vs. arousal with a default reference line (y=x)

slope and intercept parameters to `geom-abline()` or use `geom-smooth(method = "lm")` which we'll explore in our later chapter on regression analysis.

A problem that we have in our scatter plot is that many of the points are stacked on top of one another. One solution to this is to add a small amount of random noise that places some points away from each other but not enough to impact the overall interpretation of the graph. `ggplot2` makes this easy to do without changing our code. Since we are mapping the same x and y values as before, we do not need to change our call to `ggplot()`. Instead, we are merely changing the geometric objects that display our data. To create points with random noise, as shown in Fig. 4.4, we will replace our call to `geom-point()` with `geom-jitter()`:

```
# Create scatter plot with jittered points
muse |>
  filter(genre == "rap" | genre == "funk" | genre == "metal") |>
  ggplot(mapping = aes(x = valence_tags, y = arousal_tags)) +
  geom_jitter()
```

One thing to note is that if you look at code online, you will find that many people do not explicitly state the `mapping` argument when using the pipe operator. Remember in R, each function has a default order of arguments, and if you place them in the same order that R expects, you do not have to specify the argument name itself. We have already seen this in action when we passed the MuSe dataset tibble. Thus, the following will produce the same result:

```
# Create scatter plot with jittered points but without mapping argument
muse |>
  filter(genre == "rap" | genre == "funk" | genre == "metal") |>
```

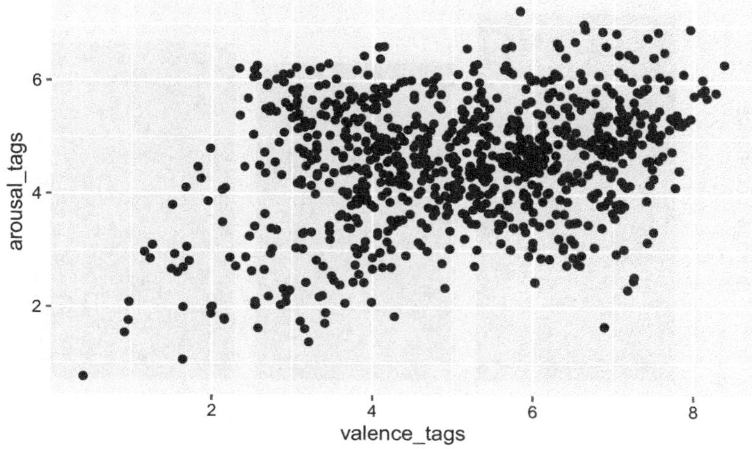

Fig. 4.4 Jittered scatter plot of valence vs. arousal to reduce overplotting

```
  ggplot(aes(x = valence_tags, y = arousal_tags)) +
  geom_jitter()
```

4.5 Creating Column Graphs

Now that we have a better understanding of how we can use the three main layers of the grammar of graphics to create a scatter plot, we can examine additional geoms and how to use them to create other plot types. Remember that in most cases, simply changing the geometric object will modify the type of graph.

Let's see how this modification of geometric objects allows us to make a column graph. We will use the same general framework we used to make our scatter plot in terms of passing our data and mapping our arguments. Column graphs require us to map at least two variables: a discrete one on the x-axis and a continuous one on the y-axis. For this example, we will look at how dominance varies by genre:

```
# Filter for specified genres and calculate average dominance by genre.
↪  Create a column graph
muse |>
  filter(genre == "rap" | genre == "funk" | genre == "metal") |>
  group_by(genre) |>
  summarize(average_dominance = mean(dominance_tags)) |>
  ggplot(aes(x = genre, y = average_dominance)) +
  geom_col()
```

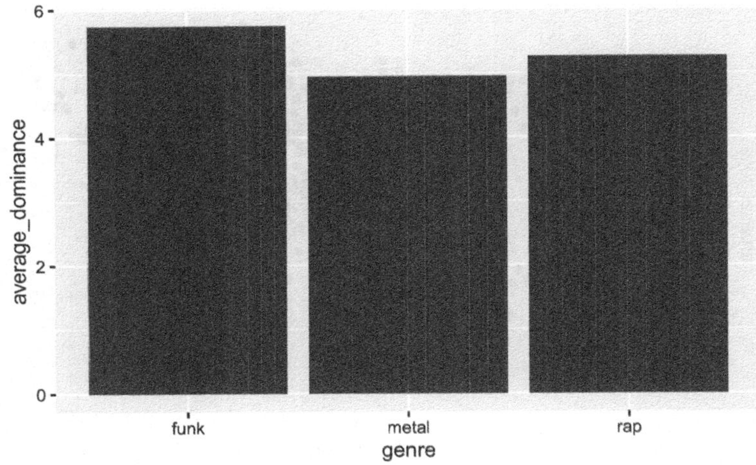

Fig. 4.5 Average dominance scores by music genre (rap, funk, metal)

Figure 4.5 highlights a fascinating variance across musical genres that indicates how different styles of music can evoke distinct levels of perceived control or empowerment. Our analysis reveals that funk music, for instance, exhibits a notably higher average value in terms of "dominance," a finding that resonates with the genre's signature musical qualities. The intrinsic upbeat tempo and compelling bass lines, often punctuated by dynamic brass sections, imbue funk with an energetic and forceful character.

Together, these insights suggest that the average "dominance" feeling is not just an abstract measurement but a tangible reflection of the music's capacity to evoke certain emotions and states of mind. As such, the "dominance" metric becomes a lens through which we can understand the impact of musical genres on our emotional landscape and how they shape and reflect cultural experiences.

4.5.1 Mapping Additional Variables

Often you want to display more than just the relationship between two variables. Whenever you are mapping a new variable to some element of your graph, you need to send it to the aesthetic mapping argument. We can add aesthetics in a myriad of ways such as changing the shape or color of each point. Let's see how this works by filling in the bar for each genre with a different color.

One thing to keep in mind is that "color" in a column or bar graph refers to the color for the outline of each column rather than inside of it. To actually change

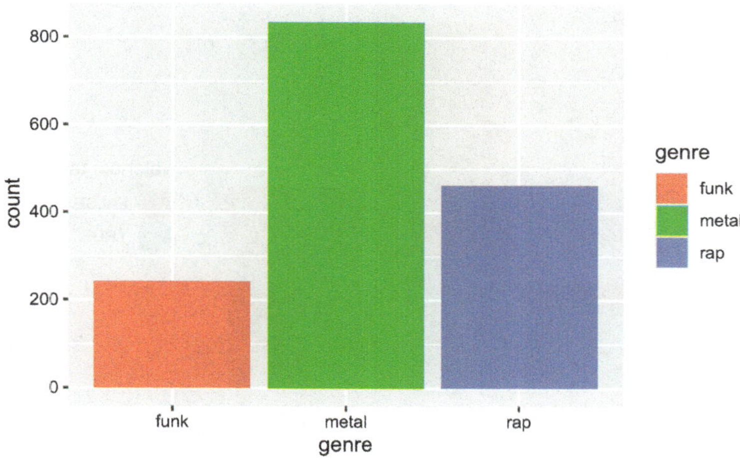

Fig. 4.6 Bar graph showing counts of songs per genre, with bars filled by genre

the color of the bars themselves, as shown in Fig. 4.6, we have to map a value to "fill":

```
# Create a bar graph with color mapped to genre
muse |>
  filter(genre == "rap" | genre == "funk" | genre == "metal") |>
  ggplot(aes(x = genre, fill = genre)) +
  geom_bar()
```

Sometimes you might want to see a shape or color based on the result of an expression. Luckily, "ggplot2" allows us to do this by mapping an expression to the aesthetic argument. R will then interpret the expression and code the points that are true as one color and the ones that are false another. Together, with the `fill` argument, this can be used to map proportions, as shown in Fig. 4.7, and we can do this using the following code:

```
# Create a bar graph with fill mapped to valence
muse |>
  filter(genre == "rap" | genre == "funk" | genre == "metal") |>
  ggplot(aes(x = genre, fill = valence_tags > 5)) +
  geom_bar()
```

As we have already seen, one of the benefits of "ggplot2" is that the knowledge you gain from working with one type of graph easily carries over to other types. Let's see how we can map the color of the points to the genre in a scatter plot, as shown in Fig. 4.8. In contrast to the bar graph, we will not use "fill" but "color." This

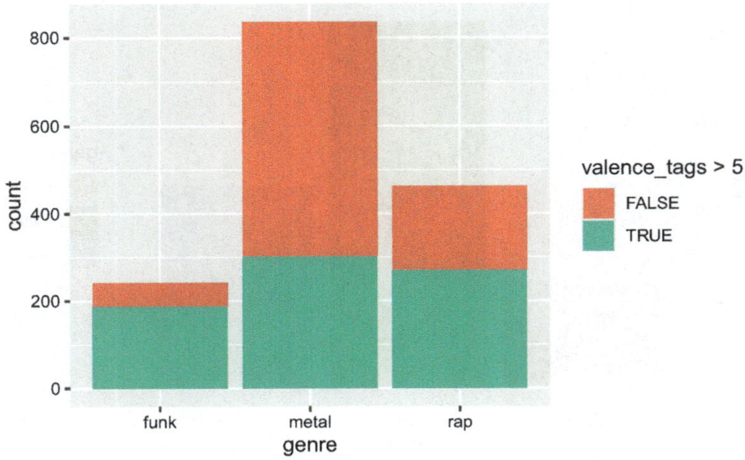

Fig. 4.7 Bar graph showing counts of songs per genre, filled based on whether valence is greater than 5

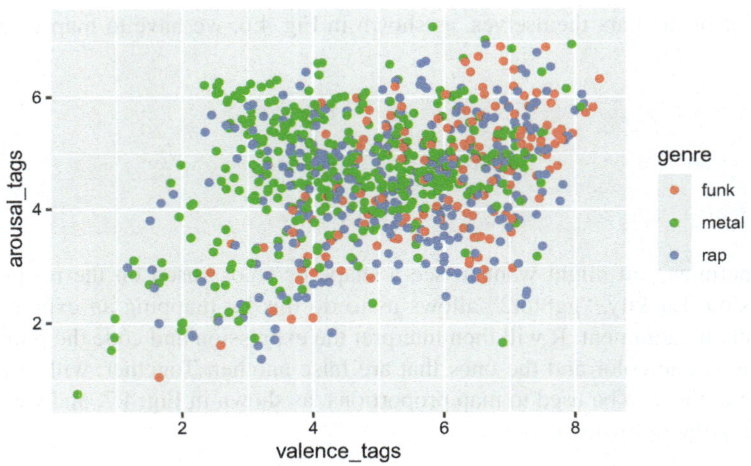

Fig. 4.8 Scatter plot of valence vs. arousal, with points colored by genre

is because points are technically only outlines and not circles. Thus, we are coloring the outline of the points rather than the circles:

```
# Create scatter plot with color mapped to genre
muse |>
  filter(genre == "rap" | genre == "funk" | genre == "metal") |>
  ggplot(aes(x = valence_tags, y = arousal_tags, color = genre)) +
  geom_point()
```

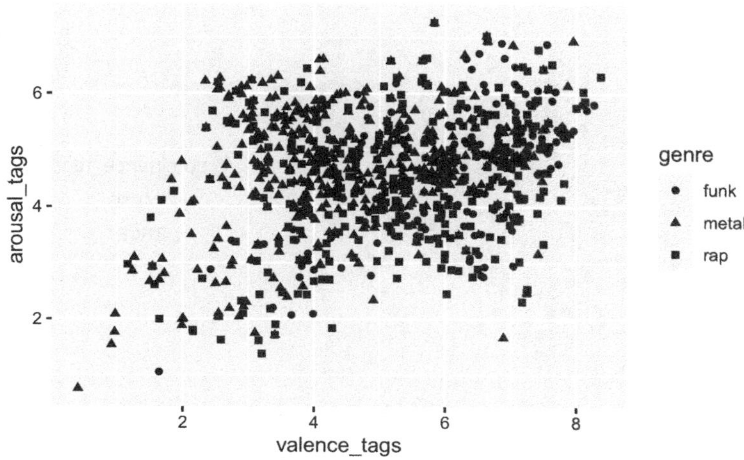

Fig. 4.9 Scatter plot of valence vs. arousal, with point shapes determined by genre

As you see, the only changes we needed to make from our earlier scatter plot was to map an additional variable to our `aes()` call that specified that we wanted color to represent genre. If we wanted to change the shape of the points based on genre, as shown in Fig. 4.9, we would replace the word "color" with "shape":

```
# Create scatter plot with shape mapped to genre
muse |>
  filter(genre == "rap" | genre == "funk" | genre == "metal") |>
  ggplot(aes(x = valence_tags, y = arousal_tags, shape = genre)) +
  geom_point()
```

Let's see how "dominance tags" varies with "arousal tags" and "valence tags." We suspect that as arousal and valence increase, the value of dominance does as well. We will pass an expression to see if dominance is greater than 5 to the color mapping in `aes()` and observe how this impacts our graph, as shown in Fig. 4.10:

```
# Create scatter plot with dominance mapped to color
muse |>
  filter(genre == "rap" | genre == "funk" | genre == "metal") |>
  ggplot(aes(x = valence_tags, y = arousal_tags, color = dominance_tags >
  ↪ 5)) +
  geom_point()
```

Figure 4.10 supports the hypothesis that dominance also goes up with valence and arousal. Our quick exploratory analyses also show us that we can broadly classify songs into two categories: those that have high values in all three categories

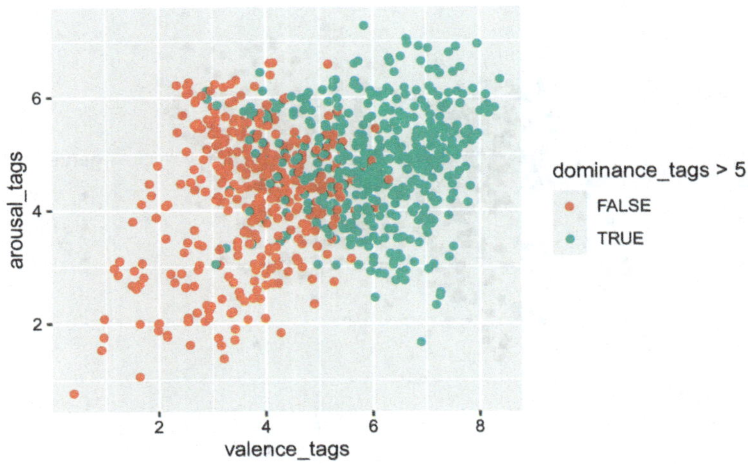

Fig. 4.10 Scatter plot of valence vs. arousal, colored based on whether dominance is greater than 5

and those that do not. Along with visualizations, we might also explore our hypotheses by doing a quick count of genre by emotional tags:

```
# Count songs by genre and emotional tags
muse |>
  filter(genre == "rap" | genre == "funk" | genre == "metal") |>
  count(arousal_tags > 3 & valence_tags > 3 & dominance_tags > 3)
```

```
# A tibble: 2 x 2
  `arousal_tags > 3 & valence_tags > 3 & dominance_tags > 3`     n
  <lgl>                                                       <int>
1 FALSE                                                         227
2 TRUE                                                         1314
```

When doing exploratory data analysis, you will find yourself jumping back and forth between strict numerical analysis and more advanced graphical methods. Use whichever method is faster and more efficient for you and/or your audience.

4.5.2 A Warning About Mapping Colors and Shapes

At the beginning of this chapter, I discussed how the grammar of graphics is less focused on aesthetic choices than on creating graphs that are fundamentally sound. We witnessed that when we chose not to map a variable directly, "ggplot2" simply used a sensible default that worked for most cases. But how exactly do we change the colors, say in a scatter plot, if we do not have a variable to map?

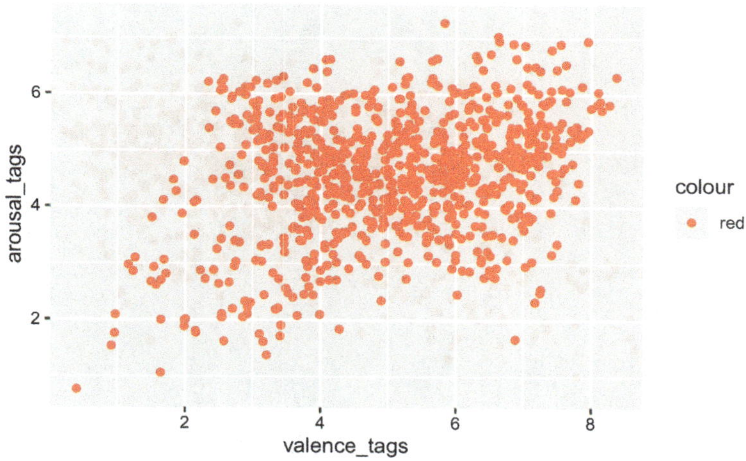

Fig. 4.11 Scatter plot of valence vs. arousal with all points set to red using aesthetic mapping

One incorrect way that you might see online is to create a field with the value we want to be mapped *inside* the `aes()` function. You saw an example of this when we took conditional expressions and mapped the results to aesthetics above. When we did that, "ggplot2" iterated through each observation and created a new Boolean value (TRUE or FALSE) based on the results of the conditional. Rather than iterating over a conditional, however, we could simply pass a static value, like "red." Let's demonstrate this approach and then show the correct way to do it.

Again, please avoid the following approach as it sidesteps the foundational principles of the grammar of graphics and can lead to unexpected results. This is only so that you can read code online and understand what is going on (Fig. 4.11).

Let's say that we are interested in setting all the points of a scatter plot to be red. To do so, we could run the following:

```
# Create scatter plot with all points red
muse |>
  filter(genre == "rap" | genre == "funk" | genre == "metal") |>
  ggplot(aes(x = valence_tags, y = arousal_tags, color = "red")) +
  geom_point()
```

Again, when we assign every observation in your dataset to the color "red," you are fabricating a variable that holds no intrinsic data value but serves as a vessel for your aesthetic preference. This makeshift field is subsequently mapped to the "color" aesthetic in "ggplot2," effectively instructing the software to render every data point in the hue of your choice, which in this case, is red. The correct approach is to set the aesthetic outside the `aes()` function, directly within the `geom_point()`

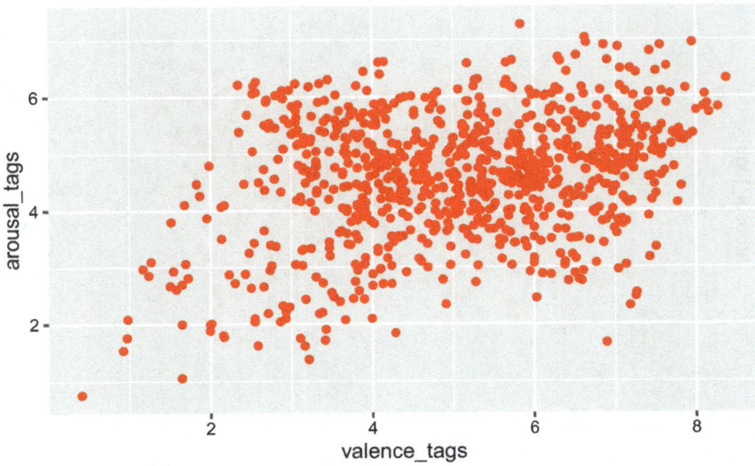

Fig. 4.12 Scatter plot of valence vs. arousal with all points correctly set to red using aesthetic setting

function. This tells `geom_point()` to simply *set* the color to red, rather than trying to map a variable (which doesn't exist) to the color aesthetic (Fig. 4.12):

```
# Create scatter plot with all points correctly set to red
muse |>
  filter(genre == "rap" | genre == "funk" | genre == "metal") |>
  ggplot(aes(x = valence_tags, y = arousal_tags)) +
  geom_point(color = "red") # This is the correct way!
```

4.5.3 How to Sort Column Graphs

A notable peculiarity when generating column graphs, particularly for those new to "ggplot2," is the default behavior regarding the arrangement of bars. They are presented in the order that the categories appear in the dataset and not by the size of the values they represent. This can lead to a visual representation that is less immediately informative, as it might obscure underlying trends or patterns that would be more apparent if the data were ordered. For instance, in a column graph depicting sales by product, we might naturally expect to see the products arranged from best to worst-selling. However, without explicit instructions to sort the data, "ggplot2" will simply plot the products in the order they are listed in the dataframe.

While "ggplot2" is a robust tool for data visualization, it introduces certain complexities when it comes to reordering column graphs. This process can be less

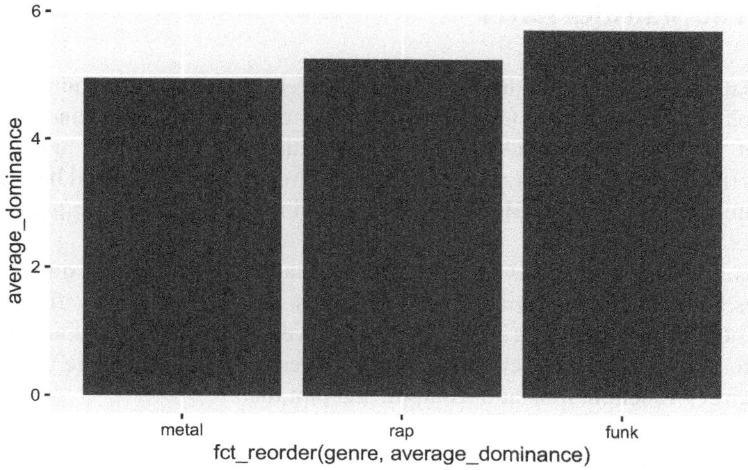

Fig. 4.13 Column graph of average dominance by genre, reordered by dominance value

than intuitive for users accustomed to other graphing tools or methods. To preserve instructional clarity and maintain a more accessible approach throughout most of this book, we will often adhere to the default order provided by "ggplot2" when creating column graphs. This choice is meant to prevent over-complication and ensure that the focus remains on grasping the core principles of the grammar of graphics.

However, understanding how to reorder factors is important for those occasions when the visual hierarchy of data points is critical for the analysis or storytelling, as demonstrated in Fig. 4.13. The `fct-reorder()` function from the "forcats" package, a member of the comprehensive "tidyverse" collection, offers a systematic way to rearrange factors by another, typically numerical, variable that dictates the desired order for the factor levels. When this reordered factor is mapped to the x-axis in "ggplot2," the categorical data is displayed in a sequence that directly corresponds to the ordering variable:

```
# Create a column graph with reordered factors
muse |>
  filter(genre == "rap" | genre == "funk" | genre == "metal") |>
  group_by(genre) |>
  summarize(average_dominance = mean(dominance_tags)) |>
  ggplot(aes(x = fct_reorder(genre, average_dominance), y =
  ↪  average_dominance)) +
  geom_col()
```

4.6 The Statistics Layer

Now that we have looked at the three required layers, we can begin examining some of "ggplot2's" other layers. One thing that often confuses both new and advanced users of "ggplot2" is the statistics layer. The "statistical" layer comes in play when the information that we want to display has not already been calculated beforehand. It is rarely called directly. However, it is vital in understanding the grammar of graphics philosophy.

Let's see how statistics works through an example. As we saw, `geom_col()` requires both an x and y variable. In many cases, x is a categorical variable and y is the count of the categories in that variable. If we were interested in knowing how many songs there are for each of our select genres, we could calculate these using the `count()` function, look at the output, and plot them:

```
# Count songs by genre
muse |>
  filter(genre == "rap" | genre == "funk" | genre == "metal") |>
  count(genre)
```

```
# A tibble: 3 x 2
  genre     n
  <chr> <int>
1 funk    242
2 metal   836
3 rap     463
```

As we can see, when we use the `count()` function, it creates a new column called "n" with the counts for each genre. Figure 4.14 shows these results graphed using `geom-col()`:

```
# Create a column graph with counts of songs by genre
muse |>
  filter(genre == "rap" | genre == "funk" | genre == "metal") |>
  count(genre) |>
  ggplot(aes(x = genre, y = n)) +
  geom_col()
```

4.6.1 *Creating Bar Graphs*

Since getting the count of a categorical variable is a common task, "ggplot2" provides us a shortcut for doing so through `geom_bar()`. Let's see what

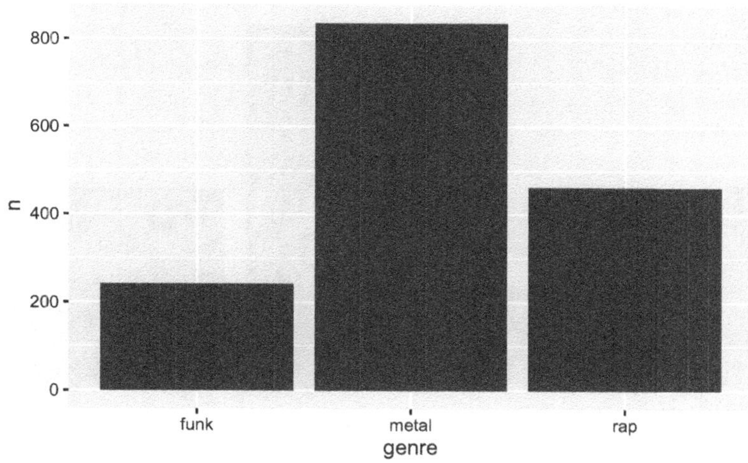

Fig. 4.14 Column graph showing the count of songs per genre, plotted from pre-calculated counts

happens when we change `geom_col()` to `geom_bar()`, which only takes an x variable:

```
# Create a bar graph with counts of songs by genre
muse |>
  filter(genre == "rap" | genre == "funk" | genre == "metal") |>
  ggplot(aes(x = genre)) +
  geom_bar()
```

As shown in Fig. 4.15, the convenience provided by “ggplot2’s” `geom-bar()` function is immediately apparent when we receive the expected graphical output without the prerequisite of manual calculations. It interprets the aesthetic mappings we provided—specifically the x variable as a categorical entity—and methodically counts the number of instances for each category. This is all done through the hidden statistics layer.

4.6.2 Statistics Layer for a Box Plot

Another common graph that calls a statistic layer is a box plot. Let’s use a box plot to see how “arousal tags” differ based on the genre. Think about what we need to create a box plot. We need to calculate the minimum, maximum, median, first quartile, and third quartile. Calculating these by hand would be time-consuming,

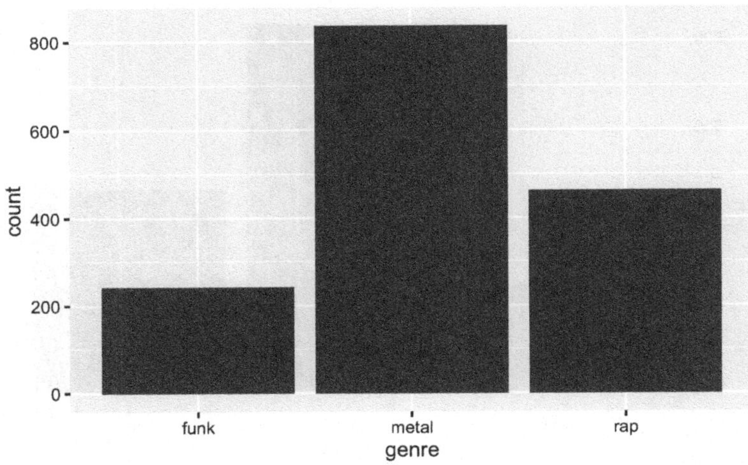

Fig. 4.15 Bar graph showing the count of songs per genre, calculated automatically by geom_bar()

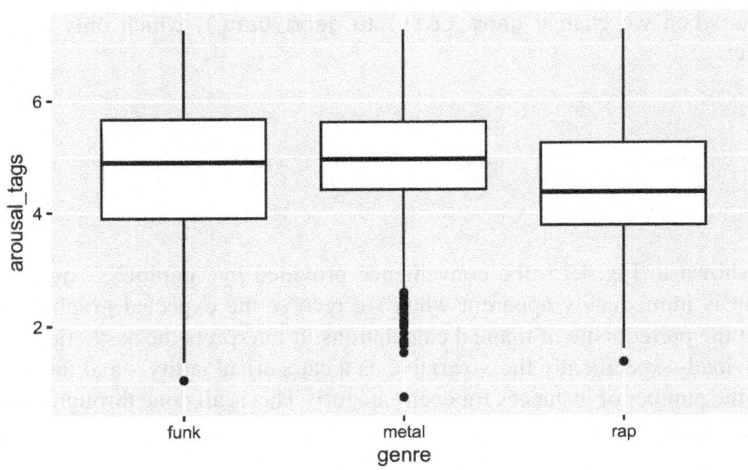

Fig. 4.16 Box plot comparing the distribution of arousal tags across different music genres

so "ggplot2" will do so automatically before displaying the necessary information using geom-boxplot(), as shown in Fig. 4.16:

```
# Create a box plot with arousal_tags by genre
muse |>
  filter(genre == "rap" | genre == "funk" | genre == "metal") |>
  ggplot(aes(x = genre, y = arousal_tags)) +
  geom_boxplot()
```

4.6.3 *The Statistics Layer for Creating Histograms*

Along with bar graphs and box plots, histograms also call a statistical function. When invoking the `geom_histogram()` function in "ggplot2," we leverage the package's built-in statistical capabilities to automatically segment the continuous variable into discrete intervals, commonly referred to as "bins." The choice of bin size and the number of bins is crucial.

By default, "ggplot2" employs a heuristic to determine an appropriate number of bins that aim to balance detail and clarity. As the package processes the variables, it calculates the frequency of data points within each bin, effectively transforming a continuous variable into a countable set of categories. These categories are then used to construct the histogram, with the height of each bar representing the number of observations that fall within the corresponding bin's range. This transformation allows us to discern patterns such as central tendencies, skewness, and the presence of multiple modes in the data, which might otherwise be difficult to detect in raw, unorganized datasets:

```
# Create a histogram with arousal_tags
muse |>
  filter(genre == "rap" | genre == "funk" | genre == "metal") |>
  ggplot(aes(x = arousal_tags)) +
  geom_histogram()
```

```
`stat_bin()` using `bins = 30`. Pick better value with `binwidth`.
```

Figure 4.17 generates a warning letting us know that we need to specify a bin value. This reminds us that the automatic binning provided by "ggplot2" may not be

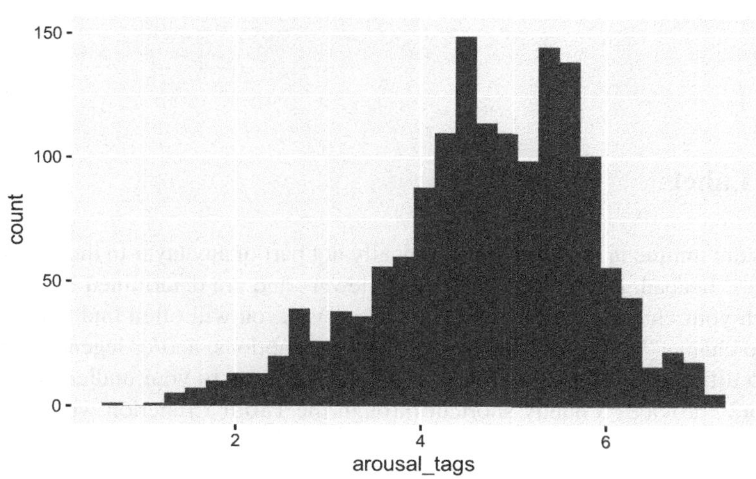

Fig. 4.17 Histogram showing the distribution of arousal tags using default binning

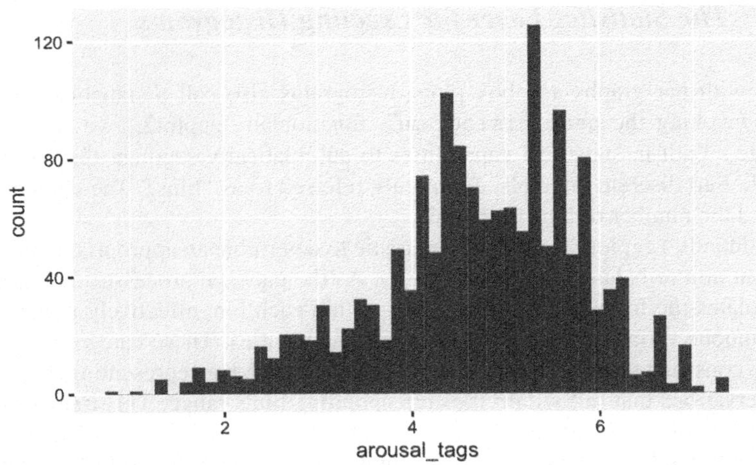

Fig. 4.18 Histogram showing the distribution of arousal tags using 50 bins

optimal for every dataset. Careful consideration and, at times, manual adjustment of binning parameters may be necessary to ensure the histogram accurately reflects the nuances of the data. The ability to specify the number of bins or the width of each bin gives the user control over the granularity of the distribution's depiction, enabling a tailored approach to data exploration and storytelling. We can do this by setting the `bins` argument in the `geom-histogram()` function, as shown in Fig. 4.18:

```
# Create histogram with 50 bins
muse |>
  filter(genre == "rap" | genre == "funk" | genre == "metal") |>
  ggplot(aes(x = arousal_tags)) +
  geom_histogram(bins = 50)
```

4.7 Labels

Labels are unique in that they are technically not part of any layer in the grammar of graphics. Instead, they are used to guide viewers and are determined by "ggplot2" through your choices in different layers. However, you will often find that you will need to change "ggplot2's" default titles, labels, captions, and/or legends all at one time to different values in order for them to make sense to your audience. Luckily, "ggplot2" provides a handy shortcut through the `labs()` function where you can

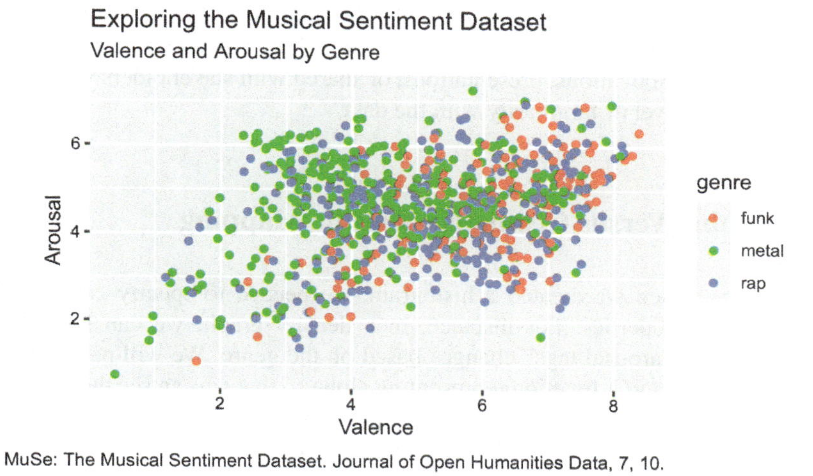

Fig. 4.19 Scatter plot of valence vs. arousal with custom title, subtitle, caption, and axis labels

pass strings as arguments specifying what values different labels will display, as shown in Fig. 4.19:

```
# Create scatter plot with labels
muse |>
  filter(genre == "rap" | genre == "funk" | genre == "metal") |>
  ggplot(aes(x = valence_tags, y = arousal_tags, color = genre)) +
  geom_point() +
  labs(
    title = "Exploring the Musical Sentiment Dataset",
    subtitle = "Valence and Arousal by Genre",
    caption = "Akiki, C., & Burghardt, M. (2021). MuSe: The Musical
    ↪  Sentiment Dataset. Journal of Open Humanities Data, 7, 10.",
    x = "Valence",
    y = "Arousal"
    )
```

For most of this book, graphs will not have labels. As with ordering, the decision to forego the explicit use of the `labs()` function in "ggplot2" for most graphs in this text is made with the intention of keeping the code as clean and concise as possible. This approach allows for a streamlined presentation that focuses on the core concepts and functionality of the package without getting caught up in the minutiae of custom labeling. The default settings within "ggplot2" are thoughtfully designed to provide clear and intelligible labels that typically suffice for the casual viewer or during initial stages of data exploration.

In situations where the default labeling might lead to ambiguity or confusion, or where a graph is intended for a wider and possibly less technical audience, taking the time to apply the `labs()` function becomes more than just a matter of aesthetics—it becomes an essential step in the responsible presentation of data. It ensures that each visual element is clearly annotated, which is especially crucial when the graph is to be included in publications, presentations, or shared with stakeholders who may not have the same level of familiarity with the data.

4.8 Global Versus Geometric-Specific Mapping

As you saw, when we created a histogram, we needed to specify certain values in geometric properties. For instance, in a density graph, we can look at how something like "arousal tags" changes based on the genre. We will pass a value to the `geom-density()` function representing alpha so that we can see the differences between the genres more clearly, as shown in Fig. 4.20. The alpha value is a number between 0 and 1, where 1 represents complete opacity:

```
# Create density plot with alpha set to 0.3
muse |>
  filter(genre == "rap" | genre == "funk" | genre == "metal") |>
  ggplot(aes(x = arousal_tags, fill = genre)) +
  geom_density(alpha = 0.3)
```

The above graph is useful for differentiating a single continuous variable. However, we run into issues when we want to look at multiple continuous variables on a single graph, such as the density for each of the three tags. How would these be mapped? Figure 4.21 addresses this by removing our aesthetic mapping from the

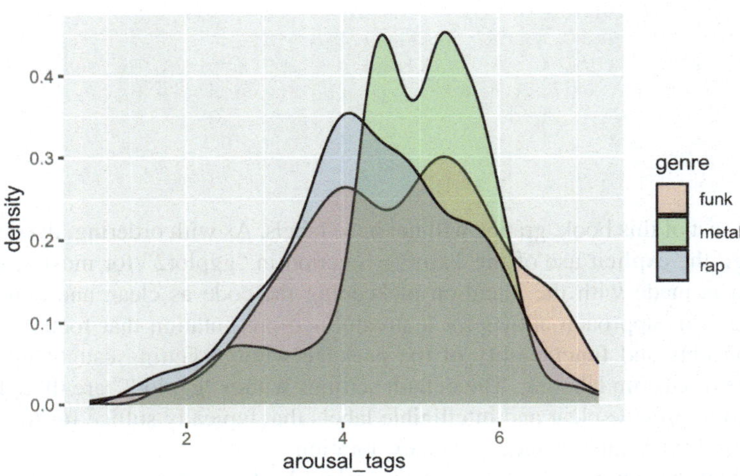

Fig. 4.20 Density plot showing the distribution of arousal tags, filled by genre with transparency

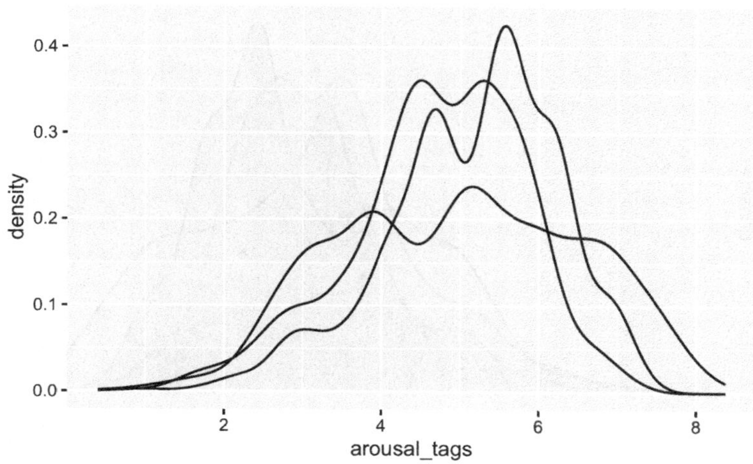

Fig. 4.21 Overlaid density plots showing distributions of arousal, dominance, and valence tags

`ggplot()` call itself since that only takes a single x variable. Instead, we will map our aesthetic values for each layer individually:

```
# Create density plot with multiple layers
muse |>
  filter(genre == "rap" | genre == "funk" | genre == "metal") |>
  ggplot() +
  geom_density(mapping = aes(x = arousal_tags)) +
  geom_density(mapping = aes(x = dominance_tags)) +
  geom_density(mapping = aes(x = valence_tags))
```

Geometric-specific mappings take precedence over those in our global `ggplot()` call. As shown in Fig. 4.22, one benefit of doing so is that you can also adjust fills, alphas, etc. for each layer:

```
# Create density plot with multiple layers and different fills
muse |>
  filter(genre == "rap" | genre == "funk" | genre == "metal") |>
  ggplot() +
  geom_density(mapping = aes(x = arousal_tags), fill = "red", alpha =
  ↪ 0.3) +
  geom_density(mapping = aes(x = dominance_tags), fill = "blue", alpha =
  ↪ 0.5) +
  geom_density(mapping = aes(x = valence_tags), fill = "green", alpha =
  ↪ 0.4)
```

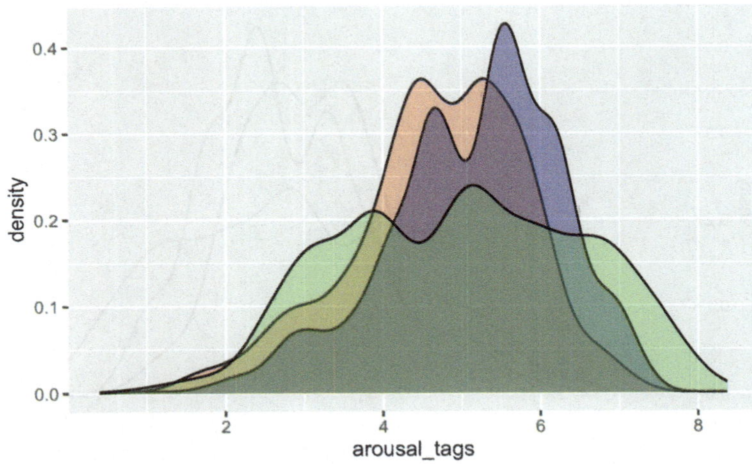

Fig. 4.22 Overlaid density plots with custom fill colors and alpha values for each tag type

4.9 Facets

Faceting is a remarkably potent feature within `ggplot2` that can exponentially increase the clarity and depth of data visualization, as demonstrated in Fig. 4.23. The concept of facets—derived from the idea of cutting many faces into a gemstone—allows for the multidimensional exploration of complex datasets. By implementing the `facet-wrap()` function, `ggplot2` presents a concise method for organizing and displaying subsets of data within individual panels. This means that a dataset can be partitioned based on a particular categorical variable, such as genre, with each category receiving its own dedicated plot, arranged in a grid for easy comparison.

The power of faceting lies in its ability to unveil patterns and trends that might be obscured when viewing aggregated data. When comparing multiple variables across different subgroups, color-coding can become overwhelming or impractical, particularly as the number of categories grows. Facets tackle this issue head-on, offering a clean, segmented view that maintains the same scale and axes across all plots. This uniformity is key as it ensures that the comparisons between subgroups are valid.

Let's see how to use facets. Instead of differentiating between genre's by color, we could add a `facet_wrap()` layer that shows each genre in its own grid:

```
# Create density plot with facets
muse |>
  filter(genre == "rap" | genre == "funk" | genre == "metal") |>
  ggplot(aes(x = valence_tags)) +
  geom_density() +
  facet_wrap(~genre)
```

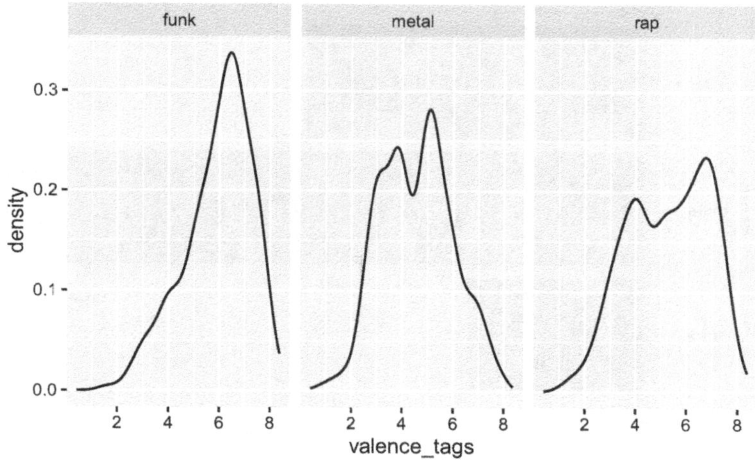

Fig. 4.23 Density plots of valence tags, faceted by music genre

The tilde (~) operator here might seem a bit unusual at first. While it's true that ~ is typically used in R to define a formula (e.g., `dependent ~ independent`), in the context of `facet_wrap()`, it serves a slightly different purpose. Here, the tilde indicates that `genre` is the variable to be used for faceting. In other words, it specifies that we want to create a separate panel for each unique value of the `genre` variable. While the tilde's usage might not be immediately obvious, it's a convention you'll quickly become familiar with when working with the "tidyverse."

4.10 Coordinates

The sixth layer of "ggplot2" is coordinates. For most graphs, you will be using the Cartesian coordinate system, and this is what "ggplot2" will default to if you do not tell it otherwise. However, there are points where we may want to flip our coordinate, such as making a horizontal bar graph. To do this, as shown in Fig. 4.24, we simply add another layer called `coord-flip()`:

```
# Create horizontal bar graph
muse |>
  filter(genre == "rap" | genre == "funk" | genre == "metal") |>
  ggplot(mapping = aes(x = genre)) +
  geom_bar() +
  coord_flip()
```

Along with flipping the coordinates, you can also use polar coordinates where x and y represent the radius and the angle, as shown in Fig. 4.25. Here is the same

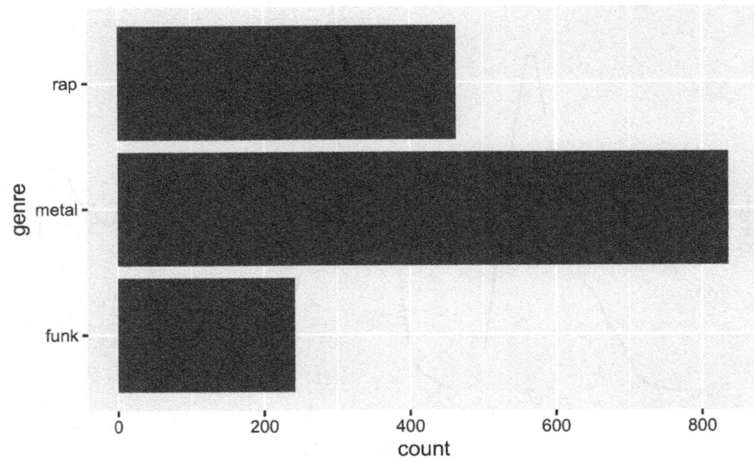

Fig. 4.24 Horizontal bar graph showing song counts per genre

Fig. 4.25 Bar graph of song counts per genre displayed using polar coordinates (wind rose plot)

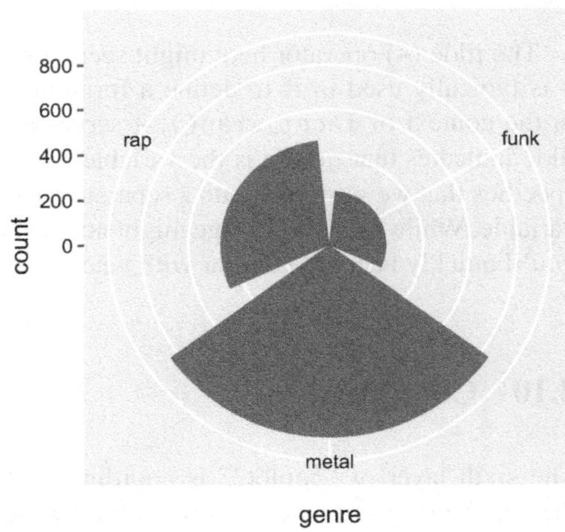

column graph as above but with polar coordinates. These types of graphs are often called "wind rose plots":

```
# Create wind rose plot
muse |>
  filter(genre == "rap" | genre == "funk" | genre == "metal") |>
  ggplot(mapping = aes(x = genre)) +
  geom_bar() +
  coord_polar()
```

Coordinates can also be used to limit the x and y labels of the graph through the `coord-cartesian()` function. For all practical purposes, you can imagine this

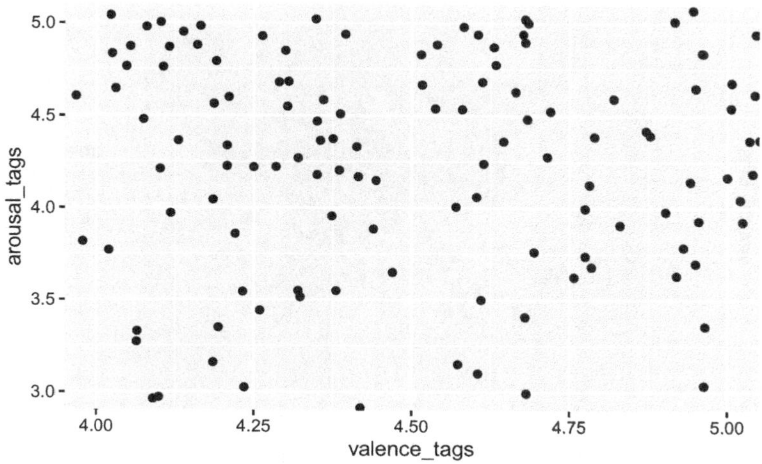

Fig. 4.26 Scatter plot of valence vs. arousal, zoomed in using `coord_cartesian()`

as "zooming in" on the graph, as demonstrated in Fig. 4.26. It takes two optional arguments that specify where the x-axis and the y-axis should be limited:

```
# Create scatter plot with limited coordinates
muse |>
  filter(genre == "rap" | genre == "funk" | genre == "metal") |>
  ggplot(aes(x = valence_tags, y = arousal_tags)) +
  geom_point() +
  coord_cartesian(xlim = c(4, 5), ylim = c(3, 5))
```

4.11 Scales

Navigating the concept of scales within the framework of the grammar of graphics can present a challenge to newcomers—it requires a slight shift in perspective from traditional plotting methods. To demystify scales, it helps to revisit the idea of mapping, which is central to how "ggplot2" creates meaningful visualizations. Mapping is the process by which we link or "map" the abstract, nonvisual data in our dataset—such as genre names—to visualize properties that can be perceived and understood like color. In many instances, you might have a categorical variable like "genre" that you wish to represent with a different color for each value.

However, data frames typically store genre names as text, not colors. This is where scaling comes into play. Scales act as the translators between the data you want to represent and the aesthetic attributes of your graph. They provide a set of rules that "ggplot2" uses to assign specific visual properties, such as colors, shapes, or sizes, to the values within your data. For example, a scale could designate "rap" to be displayed as blue, "rock" as red, and "jazz" as green. These assignments

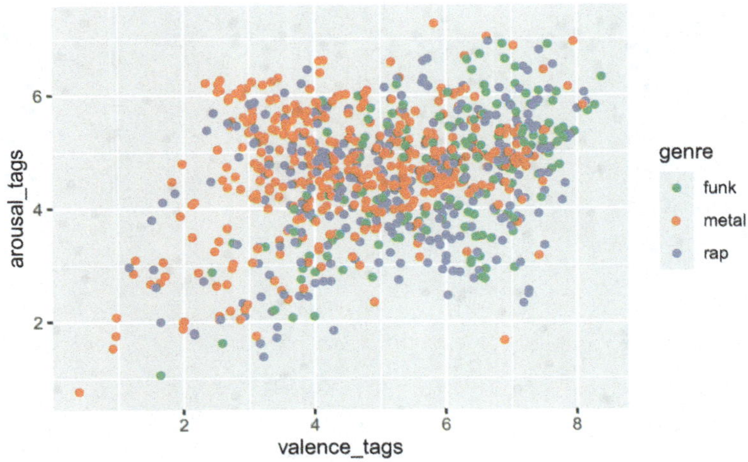

Fig. 4.27 Scatter plot of valence vs. arousal, with point colors determined by genre using the "Set2" Brewer palette

are not random; scales are crafted to maximize both the aesthetic appeal and the interpretability of the graph.

In "ggplot2," scales are automatically applied behind the scenes using a set of sensible defaults. This automatic scaling is a significant advantage—it allows users to quickly generate informative visuals without delving into the complexities of color theory or design considerations. However, as one becomes more familiar with the grammar of graphics, understanding and leveraging scales can lead to more refined and customized visual outcomes.

The ability to manually adjust scales also enables you to ensure accessibility in your graphs, such as adhering to colorblind-friendly palettes. While the default settings are suitable for exploratory analysis, the thoughtful application of scales can elevate the clarity and effectiveness of your visual communication making your data not only more visually engaging but also ensuring it is accurately and ethically represented.

We can manually change our scales if we want by using scale functions. For instance, as shown in Fig. 4.27, we can use the `scale-color-brewer()` or the `scale-fill-brewer()` with an argument representing the palette we are interested in along with a type representing if we want a sequential, differential, or diverging type of scale:

```
# Create scatter plot with color mapped to genre and color scale
muse |>
  filter(genre == "rap" | genre == "funk" | genre == "metal") |>
  ggplot(aes(x = valence_tags, y = arousal_tags, color = genre)) +
  geom_point() +
  scale_color_brewer(type = "qual", palette = "Set2")
```

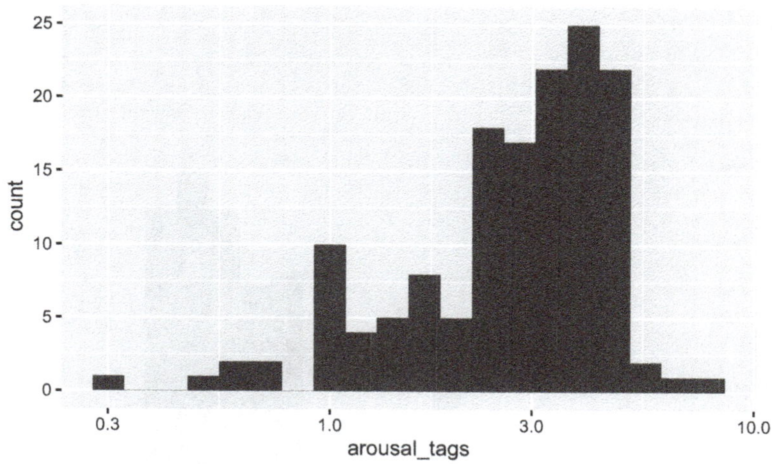

Fig. 4.28 Histogram of arousal tags for "meditation" genre, with a log10 transformed x-axis

For each aesthetic mapping, such as shape and fill, there is a corresponding scale that goes with it. However, I would recommend that you stick to using the default ones that come with `ggplot2` or your theme—which we will cover later.

We are used to thinking about the geometric objects themselves as scales, but x and y are also controlled with scales. We often need to do log transformations, for instance, let's go back and filter out "meditation" music. We know that the "arousal tags" for this are likely very tiny. As a result, as shown in Fig. 4.28, we can take the log of the x variable to adjust for right skew by using the `scale-x-continuous()` function with `"log10"` sent to the `trans` argument:

```
# Filter for genre "meditation" and create histogram with log
↪  transformation
muse |> filter(genre == "meditation") |>
  ggplot(aes(x = arousal_tags)) +
  geom_histogram(bins = 20) +
  scale_x_continuous(trans = 'log10')
```

Because these types of log transformations are common, there are shortcuts that have been created to do the same thing with `scale_x_log10()` and `scale_y_log10()`:

```
# Filter for genre "meditation" and create histogram with log
↪  transformation
muse |>
  filter(genre == "meditation") |>
```

```
ggplot(aes(x = arousal_tags)) +
geom_histogram(bins = 20) +
scale_x_log10()
```

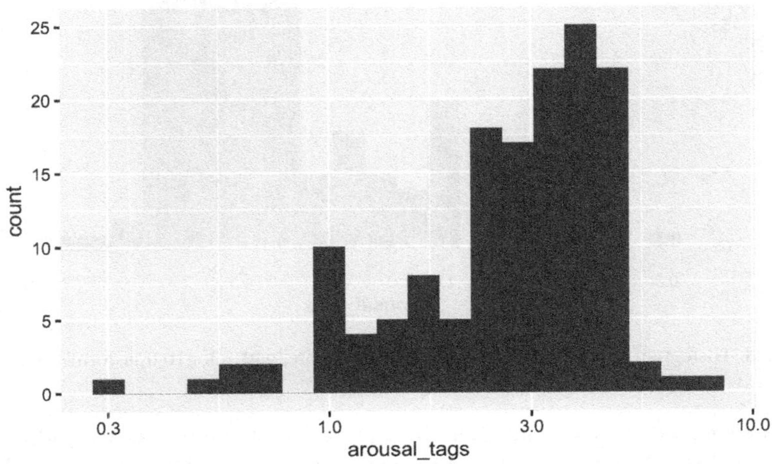

You may find that some individuals use the scale function to "zoom in" on a plot similar to the way that we used `coord_cartesian()`. Because these results resemble each other, there can be confusion on when to use each one. The problem with using scales to limit a graph is that it removes calculations on data outside the limits specified. This can be problematic in certain plots where this data is still necessary to compute. In short, use `coord_cartesian()` when trying to "zoom in" on a plot.

4.12 Themes

Although not part of Wickham's original formulation, "themes" have become a common layer for ggplot's framework. These control all the non-data-oriented ink that goes on a plot. Remember that the grammar of graphics, as formulated by both Wilkinson and Wickham, was not designed to focus on the aesthetic choices and design of the graph but to create a general framework that could create any graph. Themes allow us to make more substantial changes to how our graph looks without worrying about if our graph is mathematically sound.

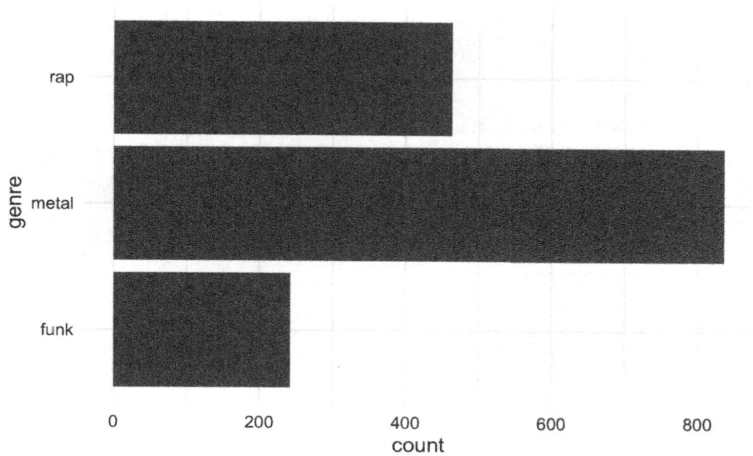

Fig. 4.29 Horizontal bar graph using theme_minimal()

"ggplot2" itself comes with a few themes built in which we can utilize by adding a layer calling theme-<theme name>(), as shown in Fig. 4.29:

```
# Create bar graph with theme_minimal
muse |>
  filter(genre == "rap" | genre == "funk" | genre == "metal") |>
  ggplot(aes(x = genre)) +
  geom_bar() +
  coord_flip() +
  theme_minimal()
```

If you find yourself using the same theme each time, it may be easier to use the theme_set() function:

```
# fig-cap: "Bar graph demonstrating what happens when you set theme to
↪  minimal for all plots."
# Set theme to minimal
theme_set(theme_minimal())

# Create bar graph with theme_minimal
muse |>
  filter(genre == "rap" | genre == "funk" | genre == "metal") |>
  ggplot(aes(x = genre)) +
  geom_bar() +
  coord_flip()
```

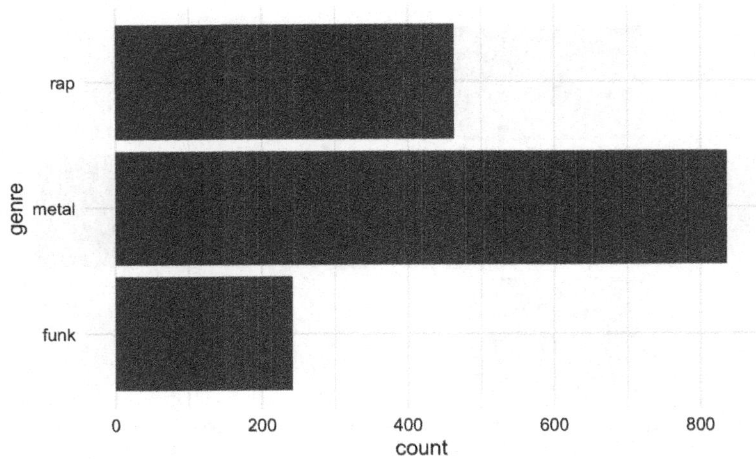

There are also additional themes that you can find online. The "ggthemes" package, for instance, provides common themes that mimic those in popular software, such as Excel, and publications, such as *The Wall Street Journal* and *The Economist* (J. Arnold 2024). You will need to install this package if you have not already before loading it. After doing so, you simply add the call to the corresponding theme as a layer. For instance, Fig. 4.30 uses the *WSJ* theme:

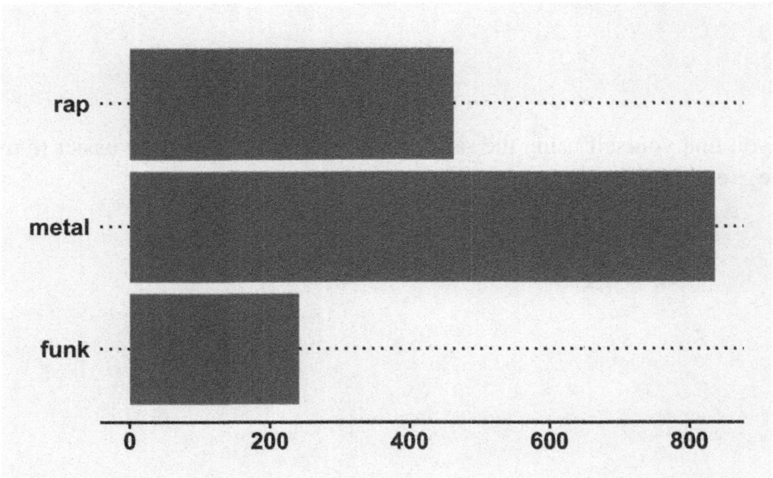

Fig. 4.30 Horizontal bar graph using the *The Wall Street Journal* theme from "ggthemes"

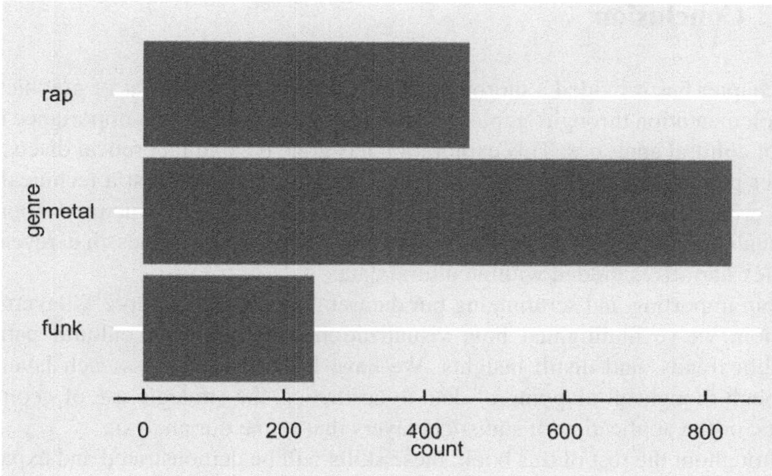

Fig. 4.31 Horizontal bar graph using *The Economist* theme from "ggthemes"

```
# Uncomment to install once
# install.packages("ggthemes")
# Load ggthemes package
library(ggthemes)

# Create bar graph with WSJ theme
muse |>
  filter(genre == "rap" | genre == "funk" | genre == "metal") |>
  ggplot(aes(x = genre)) +
  geom_bar() +
  coord_flip() +
  theme_wsj()
```

Likewise, Fig. 4.31 uses *The Economist* theme:

```
# Create bar graph with Economist theme

muse |>
  filter(genre == "rap" | genre == "funk" | genre == "metal") |>
  ggplot(aes(x = genre)) +
  geom_bar() +
  coord_flip() +
  theme_economist()
```

If you are just starting out, I recommend that you stick to one of the default themes or those that are in "ggthemes." In the future, you can tweak themes as necessary. The most important thing is to first make sure you understand all parts of the grammar of graphs before dealing with visual appeal.

4.13 Conclusion

This chapter has provided a thorough examination of the grammar of graphics and its implementation through "ggplot2" within R by emphasizing its importance in the field of cultural analytics. This exploration has gone beyond theoretical discussions to offer pragmatic insights into how data visualization is not just a technical step but a critical interpretive act. We've seen how "ggplot2's" structured approach to visualization enables the creation of compelling visual stories that reveal the complex narratives hidden within cultural data.

From importing and scrutinizing our dataset to applying "ggplot2's" layers with intention, we've highlighted how visualization can illuminate cultural patterns, highlight trends, and distill insights. We have learned to harness each layer—be it through thoughtful mapping of data to aesthetics, the strategic use of geometric objects, or the application of statistical layers that refine our analysis.

Throughout the rest of this book, these skills will be demonstrated and expanded upon through various case studies and examples. Each chapter will build on the fundamentals covered here, allowing you to apply the grammar of graphics to increasingly complex datasets and research questions.

Chapter 5
Analyzing Text the Tidy Way

Abstract The rapidly expanding field of cultural analytics has necessitated the development of robust methods for analyzing large volumes of textual data. Central to this endeavor is transforming text, a fundamentally human medium, into quantitative data that machines can process. This chapter details this transformation by applying the tidy text format, simplifying complex text into manageable tokens. Techniques such as sentiment analysis and term frequency-inverse document frequency (TF-IDF) probe deeper into the language, revealing patterns that indicate overall sentiment and word significance respective to distinct contexts. The chapter culminates in exploring topic modeling, specifically latent Dirichlet allocation (LDA), to discern thematic structures within a corpus of text by Mary Shelley, F. Scott Fitzgerald, and Charlotte Perkins Gilman. This analytical process highlights the power of tidy text in unveiling cultural and linguistic patterns and the potential of computational methods in enriching our understanding of cultural phenomena through text.

Keywords Text analysis · Tidy text · Sentiment analysis · Tokenization · TF-IDF · Topic modeling · Latent Dirichlet allocation · Computational methods · Linguistic patterns

The field of cultural analytics is rich with data that predominantly comes in the form of text—from classic literature and contemporary articles to social media posts and historical documents. Analyzing this type of data, however, presents unique challenges, since computers inherently do not "understand" human language. Consequently, to bridge this gap, we must translate our text into a form that machines can process: numbers. But how exactly do we do this?

David Robinson and Julia Silge advocate for a structured approach to text analysis known as "tidy text." They argue that by organizing textual data into a standard, clean format—specifically, a table where each row contains a single "token" (a meaningful unit of text like a word, phrase, or even emoji)—we can more efficiently and effectively apply analytical methods to it (Silge and Robinson 2017). This format is particularly beneficial because it enables a wide range of

N. Siddiqui, *Cultural Analytics in R: A Tidy Approach*, Use R!,
https://doi.org/10.1007/978-3-031-96618-7_5

manipulations and transformations that we have already seen with the "tidyverse." Additionally, it also makes it easier to jump to more advanced standardized formats, such as the text interchange format (T. Arnold et al. 2023).

Tidy text draws on a "bag-of-words" approach where each word is viewed as an independent unit. While techniques utilizing vector models and neural networks have gained popularity in other fields, a lot of cultural analytics continues to rely on bag-of-words models (Matthew L. Jockers and Thalken 2020). When utilizing this methodology, the main focus is on the frequency or presence of tokens. Depending on the research question, scholars may perform various transformations on these raw tokens before analyzing and interpreting the result (HaCohen-Kerner et al. 2020).

If you are coming from a traditional humanities and/or social science background, one of the more jarring elements of using bag-of-words is its discarding of the grammatical structure of the text. In other words, the sentence "the dog ran outside," with its clear subject-verb-object order, is rendered indistinguishable from a jumbled "outside the ran dog." The poetic cadence of a Shakespearean sonnet, the strategic emphasis of a political speech, or the layered clauses of a complex legal document are all flattened into a list of words. Nonetheless, cultural analytic researchers have shown that even when simplifying text to its barest components, we can still shed considerable insight into linguistic usage phenomena (Underwood 2019).

A useful analogy for understanding this methodology is the game *Scrabble*. Imagine that each tile in the game carries not just a single letter, but entire words—each word representing a fragment of a larger narrative. In this game, the bag of tiles becomes a vessel for the complexity and variety of language found within the body of text. When we reach into the bag and extract a tile, we're not only bringing a word into focus but also unveiling the overarching story told one word at a time.

These tiles, much like the tokens in our bag-of-words analysis, represent the essential building blocks of text. As we draw each tile, or "token," and lay it out before us, we simultaneously take note of its distinct characteristics—the way one might observe the color or thickness of a physical Scrabble tile. In text analysis, these characteristics can include metadata such as the author's name, the genre, the publication date, or even the specific chapter or verse from which the word is drawn.

As mentioned, tokenization does not have to stop at the level of individual words. Our linguistic Scrabble game can be played with a variety of different tile sizes: perhaps some tiles are bigrams, capturing word pairs that often travel together, like "New York" or "climate change," while others might be entire sentences.

In this chapter, we will delve into the intricate process of transforming unstructured text into tidy text using the "tidytext" package developed by Silge and Robinson (Queiroz et al. 2023). The "tidytext" package provides a robust framework for working with textual data in a structured and efficient manner. It facilitates the conversion of raw text into a tidy format, where each row represents a single word or token, allowing for easy analysis and manipulation.

To illustrate this process, we will analyze a corpus of literary works by Charlotte Perkins Gilman, F. Scott Fitzgerald, and Mary Shelley downloaded from Project

Gutenberg. These authors have made significant contributions to their respective genres, and their texts provide rich material for both qualitative and quantitative analysis. By applying the "tidytext" approach, we can explore the nuances of their writing, identify common themes and motifs, and uncover patterns in their word choices.

5.1 Reading in the Literature Dataset

Before we can begin our analysis, we need to load the dataset containing the literary works of Charlotte Perkins Gilman, F. Scott Fitzgerald, and Mary Shelley. The dataset consists of the complete texts of several of their works available on Project Gutenberg. Along with the text, Project Gutenberg often contains additional elements about illustrations, notes, and editorial content. For the purpose of our analysis, we will leave these elements in the dataset, but it is important to note that a more thorough examination should consider whether to remove these elements, since, while they provide valuable context, they may also introduce noise into the analysis.

Just as before, we can use the `read_csv()` function from the "readr" package to read in the dataset. Let's load the dataset and take a look at the first few rows to understand its structure. Make sure that you have the "tidyverse" package loaded before running the code below:

```
# Read the books data from the CSV file
books <- read_csv("data/ch4_books.csv")
```

As we can see, our dataset only contains three columns:

title: The title of the book
author: The author of the book
text: The text of the book as a single character string

This data will allow us to answer numerous research questions. For instance, by looking at lexical relationships and co-occurrences that may not be immediately apparent, we can begin to piece together common themes and shared experiences. Does the word "monster" appear more frequently in Mary Shelley's "Frankenstein" than in the works of F. Scott Fitzgerald or Charlotte Perkins Gilman? Are there particular words that are unique to each author, or do they share a common vocabulary? These are just a few of the questions that we can explore through text analysis. What is the sentiment of each text? Is F. Scott Fitzgerald's writing more positive or negative compared to Mary Shelley's?

5.2 Preprocessing Text Data

To make this textual information computationally tractable and amenable to systematic analysis, we will need to convert it to the "tidy text" format. This involves breaking down the text into individual tokens (words or phrases) and organizing them into a structured format that can be easily manipulated and analyzed. When you have a large corpus of text, this process can be quite complex, but the "tidytext" package provides a set of functions that make this task more manageable.

Rather than doing this manually, we can use the `unnest_tokens()` function from the "tidytext" package, which will automatically break down the text so that each row contains a single token. Doing this will simplify our ability to conduct manipulation tasks such as filtering, grouping, and summarizing. The function takes three arguments. The first argument specifies the name of the new column that will store the output—the individual tokens derived from the text transformation. We will name this column "word" to indicate our intent to analyze the text at this granularity. The second argument directs the function to the specific column in our dataset that contains the full body of text we wish to tokenize—"text" in our context. The third argument delineates the nature of the tokenization process—essentially, what shape these tokens will take. The token type can vary significantly depending on our analytical goals. While the default setting is to tokenize by individual words, the function offers flexibility to accommodate n-grams, sentences, or entire paragraphs as tokens, providing a versatile toolkit.

If you have not already done so, make sure to install and load the "tidytext" package before running the code below:

```
# Load the tidytext package
# install.packages("tidytext")
# library(tidytext)

# Tokenize the text data
books_tokens <- books |>
  unnest_tokens(word, text)
```

Let's use `glimpse()` to take a look at the structure of the `books_tokens` dataset to see how the text has been transformed into individual tokens:

```
# Display the structure of the books_tokens dataset
glimpse(books_tokens)
```

```
Rows: 1,547,390
Columns: 3
$ title  <chr> "All the Sad Young Men", "All the Sad Young Men", "All the Sad ~
$ author <chr> "F. Scott Fitzgerald", "F. Scott Fitzgerald", "F. Scott Fitzger~
$ word   <chr> "all", "the", "sad", "young", "men", "_by_", "f", "scott", "fit~
```

As we can see, the "text" column has been transformed into a new column called "word," where the text has been made lowercase and punctuation has been removed. This transformation ensures that each word is standardized, facilitating more accurate analyses by reducing variability caused by different capitalizations or extraneous punctuation. Additionally, the columns "title" and "author" have been retained for each row, allowing us to maintain the context of each token within the original text. By preserving these contextual elements, we can perform more nuanced analyses that consider the source and authorship of each word, which can be particularly useful in comparative studies or when attributing stylistic features to specific authors.

While this format may take some getting used to, once you do, you'll find that it offers a wealth of possibilities for exploring the structure and content of text. In the next section, we will delve into some of the analytical techniques that we can apply to this tidy text data to uncover patterns and insights that may not be immediately apparent from the raw text.

5.2.1 Default Tidy Text Transformations

You may have noticed that the `unnest_tokens()` function automatically applies certain transformations to the text. The removal of punctuation is one such change, which serves to eliminate extraneous characters that could interfere with the analysis. Punctuation marks often don't carry meaning on their own when considering word frequency and sentiment analysis. As such, stripping them away helps to focus on the textual content that holds semantic weight.

Another standard modification implemented during tokenization is the conversion of all text to lowercase. This ensures that words are recognized as the same token regardless of their place in a sentence, which is crucial for accurate word frequency counts. For example, without this step, "Happy," "happy," and "HAPPY" would be treated as distinct tokens, leading to fragmented and unreliable data.

While these defaults facilitate a more streamlined analysis, they may not suit all research needs (Uysal and Gunal 2014). In certain contexts, the case of letters might bear significance. For instance, in sentiment analysis, capitalization can convey emphasis and intensity of emotion, which could be a valuable aspect of the data to preserve. Similarly, in named entity recognition tasks, proper nouns are typically capitalized, and retaining this information could be critical. The `unnest_tokens()` function includes the `to_lower` argument, which you can set to `FALSE`, to opt out of these transformations.

5.2.2 Longer Tokens

Although we will focus on tokens of single words, it is helpful to have a foundation on how to split text into other types of tokens as well. In this section, we will briefly explore tokenizing into sentences and n-grams, but you should consult the help function of `unnest_tokens()` to see all options.

You are probably familiar with a sentence, but unless you have a background in natural language processing, the term n-gram may be unfamiliar. An n-gram is simply a contiguous sequence of "n" items from a given sample of text or speech. The "n" stands for a number indicating the number of elements in that textual sequence. For instance, when we talk about a bigram (i.e., an n-gram of 2), we are referring to a sequence of two consecutive words, such as "the dog" or "ran outside." A trigram (i.e., an n-gram of 3) would be a sequence of three words, like "the dog ran" or "dog ran outside." The concept of n-grams is particularly useful in natural language processing tasks such as language modeling, machine translation, and text generation. Because an n-gram can be any arbitrary number of values, you will need to pass the number of words to the `n` argument when using the `unnest_tokens()` function.

Below is a code that demonstrates the process of tokenizing our original book data into sentences and bigrams:

```
# Tokenize the text data into sentences
books_sentences <- books |>
  unnest_tokens(sentence, text, token = "sentences")

# Tokenize the text data into bigrams
books_bigrams <- books |>
  unnest_tokens(bigram, text, token = "ngrams", n = 2)
```

Just as before, we can use the `glimpse()` function to examine the structure of the `books_sentences` and `books_bigrams` datasets to see how the text has been transformed into sentences and bigrams, respectively:

```
# Display the structure of the books_sentences dataset
glimpse(books_sentences)
```

```
Rows: 82,264
Columns: 3
$ title    <chr> "All the Sad Young Men", "All the Sad Young Men", "All the Sa~
$ author   <chr> "F. Scott Fitzgerald", "F. Scott Fitzgerald", "F. Scott Fitzg~
$ sentence <chr> "all the sad young men     _by_  f.", "scott fitzgerald     n~
```

```
# Display the structure of the books_bigrams dataset
glimpse(books_bigrams)
```

```
Rows: 1,547,368
Columns: 3
$ title  <chr> "All the Sad Young Men", "All the Sad Young Men", "All the Sad ~
$ author <chr> "F. Scott Fitzgerald", "F. Scott Fitzgerald", "F. Scott Fitzger~
$ bigram <chr> "all the", "the sad", "sad young", "young men", "men _by_", "_b~
```

Just as with individual words, the `unnest_tokens()` function has stripped away punctuation and converted the text to lowercase in the sentence and bigram datasets. Why would you want to tokenize text into sentences or bigrams? Although words are the most common unit of analysis in text mining, there are situations where analyzing text at a higher level of granularity can provide valuable insights. For instance, by converting text into sentences, we can gain a higher-level view of the text, allowing us to analyze the flow and coherence of the writing. Bigrams, on the other hand, capture the relationships between adjacent words, revealing patterns and associations that may not be immediately apparent from individual words.

For the rest of the chapter, we will focus on analyzing individual words, but keep in mind that these other tokenization methods are available to you and can be useful depending on the research questions you are exploring.

5.3 Token Count and Stop Words

One of the most common starting points in text analysis is to count the frequency of each word or the set of words in our dataset. This can be done using the `count()` function, which will effectively generate a term frequency table. The output of the `count()` function is automatically arranged into two columns: one for the "word" itself and a new field labeled "n," which represents the frequency of each word's occurrence within the dataset. We will also sort the words in descending order of frequency to make it easier to identify the most common terms:

```
# Count the frequency of each word in the dataset
word_counts <- books_tokens |>
  count(word, sort = TRUE)

# Display the first few rows of the word_counts dataset
head(word_counts)
```

```
# A tibble: 6 x 2
  word      n
  <chr> <int>
1 the   84301
2 and   53033
3 of    51755
4 to    43480
5 a     31826
6 in    25086
```

The resulting table provides a snapshot of the most frequently occurring words in the dataset, which can be a valuable starting point for further analysis. However, not all words are equally informative or relevant to our research questions. Common words such as "the," "and," or "of" are ubiquitous in the English language and may not contribute much to our understanding of the text. These words are known as "stop words" and are often removed from the analysis to focus on more meaningful terms.

Several R packages provide convenient functions to automate the process of filtering out stop words. However, the "tidytext" package employs a different strategy: instead of a built-in function, it uses anti-joins. We discussed anti-joins briefly in Chap. 2. Unlike traditional joins that try to keep certain rows when they are in both tables, anti-joins remove those rows that are shared between datasets. In our example, the anti-join will look at the "word" field in the books_tokens dataset and compare it to the "word" field in the stop_words dataset. If a word is found in both datasets, it will be removed from the books_tokens dataset. This process effectively filters out the stop words from our text data.

Rather than creating our own list of stop words, we can use the `get_stopwords()` function within "tidytext" to get a dataset containing a predefined list of stop words. By executing this function without any arguments, you will receive the default stop words set, which is the English stop words derived from the "snowball" lexicon—a comprehensive collection that is widely used in text processing applications. However, text data can also be multilingual, and the need to access stop words in different languages can arise. To accommodate this, `get_stopwords()` allows us to specify a two-letter ISO language code. For example, by passing "es" as an argument, we would retrieve a list of stop words in Spanish.

For now, we will focus on the English stop words. Let's go ahead and look at some of the words in the default stop words dataset by calling the function. Note that the first time you run this command, you may need to install the `stopwords` package:

```
# Install the stopwords package if you haven't already
# install.packages("stopwords")

# Get the default English stop words
get_stopwords()
```

```
# A tibble: 175 x 2
   word      lexicon
   <chr>     <chr>
 1 i         snowball
 2 me        snowball
 3 my        snowball
 4 myself    snowball
 5 we        snowball
 6 our       snowball
 7 ours      snowball
 8 ourselves snowball
 9 you       snowball
```

```
10 your      snowball
# i 165 more rows
```

Upon inspection of the stop words list, you will see that there is considerable overlap with the most frequently occurring words from our earlier term frequency analysis. We can now perform an anti-join using the `anti_join()` function from "dplyr," passing our dataset and the stop words dataset to filter these out. During this operation, any words in our dataset that match entries in the stop words list are excluded from the resulting output:

```
# Filter out the stop words from the dataset
books_tokens <- books_tokens |>
  anti_join(get_stopwords())

# Glimpse the structure of the updated dataset
glimpse(books_tokens)
```

```
Rows: 760,595
Columns: 3
$ title  <chr> "All the Sad Young Men", "All the Sad Young Men", "All the Sad ~
$ author <chr> "F. Scott Fitzgerald", "F. Scott Fitzgerald", "F. Scott Fitzger~
$ word   <chr> "sad", "young", "men", "_by_", "f", "scott", "fitzgerald", "new~
```

Having filtered out the top stop words using the anti-join operation, we can now proceed with our analysis. Let's recompute the word frequency counts and examine the most common words in our dataset:

```
# Recompute the word frequency counts
word_counts <- books_tokens |>
  count(word, sort = TRUE)

# Display the first ten rows of the word_counts dataset
head(word_counts, 10)
```

```
# A tibble: 10 x 2
   word       n
   <chr>  <int>
 1 one     5024
 2 said    3360
 3 now     3141
 4 life    2612
 5 man     2282
 6 like    2246
 7 time    2161
 8 little  2155
 9 yet     1999
10 see     1998
```

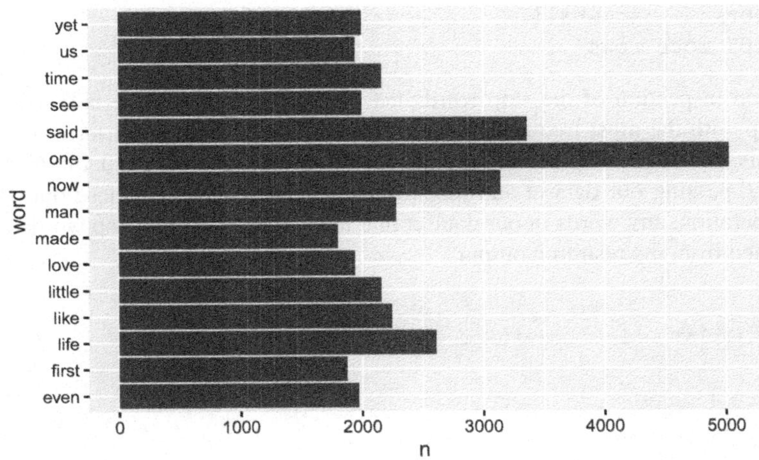

Fig. 5.1 Top 15 most frequent words in the corpus after removing default stop words

Our top words are now the following: one, said, now, life, man, like, time, little, yet, and see. The authors we are looking at often utilized themes related to individual experiences, existential reflections, and social observations. Words such as "one," "life," and "man" reflect the frequent focus on individual characters and their conditions. Similarly, verbs like "said" and "see" indicate a narrative style that involves dialogue and direct descriptions of actions and perceptions. Words like "now," "yet," and "like" are indicative of temporal and comparative contexts commonly found in literary prose.

Having filtered out the top stop words using the anti-join operation, we can now use "ggplot2" to visualize our word frequency counts, as shown in Fig. 5.1. These words, which have survived the culling process, presumably carry with them the essence of the subject matter at hand, and their frequency of occurrence is of particular interest to us. We will create a bar graph and focus on the top 15 words:

```
# Create a horizontal column graph of the top 15 words
word_counts |>
  top_n(15) |>
  ggplot(aes(x=word, y=n)) +
  geom_col() +
  coord_flip()
```

Even though we have removed the stop words, we can see that the most common words are still quite general and may not provide much insight into the content of the texts. We may want to consider additional filtering to focus on more specific terms that are likely to be more informative. One way to do this is to filter out words that are too short or too long, as these are often less meaningful in the context of text analysis. We can use the `filter()` function from "dplyr" to remove words that are less than three characters or more than 15 characters in length. For this, we can use the `nchar()` function, which returns the number of characters in a string.

We can also consider removing words that we ourselves deem uninformative or irrelevant to our analysis. For example, we may want to remove "said" or "see" from our analysis. Although they were not included in the default stop words list, they may not be particularly informative. We can do this by using the `filter()` function from "dplyr" to exclude words that meet these criteria.

Furthermore, we can use regex to remove words that contain numbers. This can be useful if we are not interested in numerical values or references in our analysis. We can use the `grepl()` function to identify words that contain numbers and then filter them out. An overview of regex is beyond the scope of this chapter, but it is a powerful tool for pattern matching and text manipulation. It allows us to search for specific patterns within text data and perform operations based on those patterns. The pattern below will match any word that contains a number:

```
# Filter out words that are less than three characters or more than 15
↪   characters and remove specific words
books_tokens <- books_tokens |>
  filter(nchar(word) >= 3, nchar(word) <= 15) |>
  filter(!word %in% c("said", "see")) |>
  filter(!grepl("[0-9]", word))

# Recompute the word frequency counts
word_counts <- books_tokens |>
  count(word, sort = TRUE)

# Get the top 10 words
head(word_counts, 10)
```

```
# A tibble: 10 x 2
   word       n
   <chr>  <int>
 1 one     5024
 2 now     3141
 3 life    2612
 4 man     2282
 5 like    2246
 6 time    2161
 7 little  2155
 8 yet     1999
 9 even    1970
10 love    1941
```

We now see that "love" is the tenth most common word. Perhaps the authors we are analyzing have a predilection for themes of love and relationships. Figure 5.2

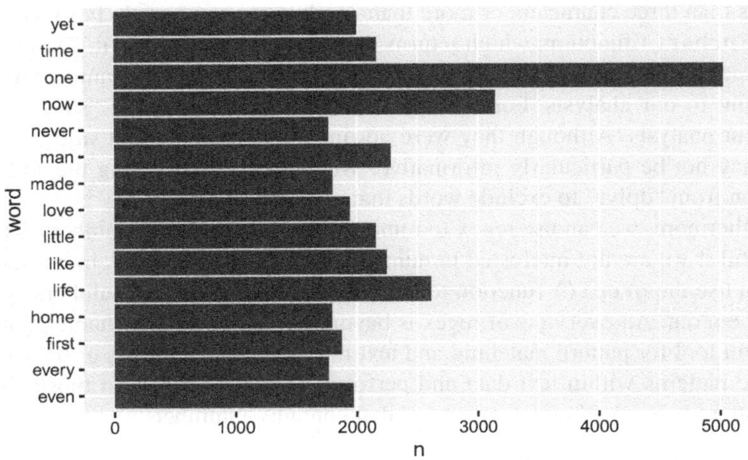

Fig. 5.2 Top 15 most frequent words after additional filtering (length, specific words, numbers)

visualizes the updated word frequency counts to see how the removal of short and long words, as well as specific words, has affected the distribution of terms:

```
# Create a column graph of the top 15 words
word_counts |>
   top_n(15) |>
   ggplot(aes(x=word, y=n)) +
   geom_col() +
   coord_flip()
```

The removal of short and long words, as well as specific words, has led to a more focused set of terms that are likely to be more informative for our analysis. However, we have only analyzed the word frequency for the entire corpus of texts. Rather than looking at the words collectively, we can also analyze them individually to see how the word frequencies differ between authors. This can provide insights into the unique writing styles and themes of each author. We will use the `group_by()` function from "dplyr" to group the data by author and then compute the word frequency counts for each author:

```
# Compute the word frequency counts for each author
author_word_counts <- books_tokens |>
  group_by(author, word) |>
  count(sort = TRUE)
```

If we take a look at the `author_word_counts` dataset, we will see that we get a count of each word for each author. However, when we sort the data, we see that multiple authors show up in the top ten. This is because we still need to sort the data by author and then by count. We can do this by using the `arrange()` function from "dplyr." Since we only want the top words for each author, we will also use the `top_n()` function from "dplyr." Let's take a look at the top four words for now to make it easier to see the data. However, you can easily change the number of words you want to see by changing the number in the `top_n()` function:

```
# Select the top ten words for each author
top_words_per_author <- author_word_counts |>
  group_by(author) |>
  top_n(4, n) |>
  arrange(author, desc(n))

# Display the top words for each author
top_words_per_author
```

```
# A tibble: 12 x 3
# Groups:   author [3]
   author                     word        n
   <chr>                      <chr>   <int>
 1 Charlotte Perkins Gilman one       1388
 2 Charlotte Perkins Gilman home      1286
 3 Charlotte Perkins Gilman women     1073
 4 Charlotte Perkins Gilman life       876
 5 F. Scott Fitzgerald      one       1248
 6 F. Scott Fitzgerald      anthony    852
 7 F. Scott Fitzgerald      amory      838
 8 F. Scott Fitzgerald      like       827
 9 Mary Shelley             one       2388
10 Mary Shelley             now       1841
11 Mary Shelley             yet       1487
12 Mary Shelley             heart     1362
```

We already see some interesting patterns emerge. For instance, Charlotte Perkins Gilman's top words—"one," "home," "women," and "life"—reflect her focus on domestic spaces and gender issues. In "The Yellow Wallpaper," arguably her most famous work, the protagonist's confinement to her home is a central theme. In contrast, F. Scott Fitzgerald's frequently occurring words highlight his character-driven narratives and descriptive style. Names like "Anthony" and "Amory" point to central characters in his novels, emphasizing character development.

Figure 5.3 attempts to graphically represent these top words for each author using a faceted bar plot that displays the top words for each author side by side. This visualization will allow us to compare the most common words used by each author and identify any distinctive patterns or themes that emerge:

```
# Create a faceted bar plot of the top words for each author
author_word_counts |>
  top_n(15) |>
```

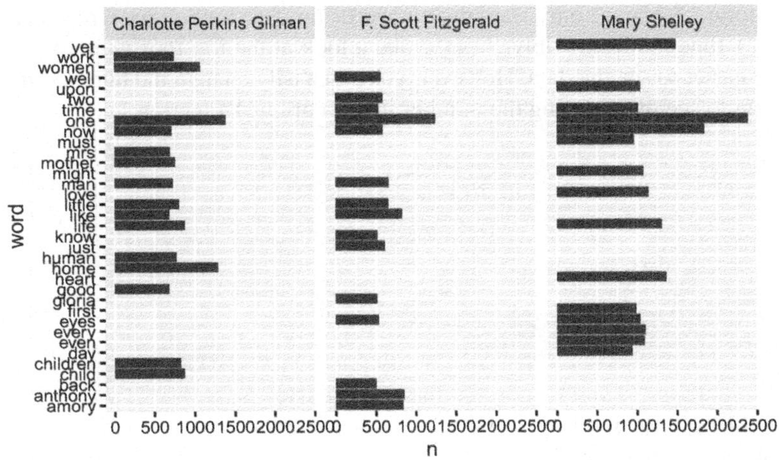

Fig. 5.3 Top words per author, faceted (note inconsistent y-axis scaling)

```
group_by(author) |>
top_n(15, n) |>
ggplot(aes(x=word, y=n)) +
geom_col() +
coord_flip() +
facet_wrap(~author)
```

Although the initial objective was to filter and present the top 15 words used by each author, a closer examination reveals the existence of more than 15 words. What happened? The reason for this unexpected discrepancy lies in the dynamics of faceted graphing. When using `facet_wrap()` to create individual plots for each branch within the same graph, `ggplot2` maintains consistency across all facets by default. This means that each individual graph within the faceted plot uses the same scales for both the x- and y-axes.

As a result, when we get the top 15 words within each author, we actually end up with the combined set of unique words that appear in the top 15 of any author. If there is minimal overlap in the top words between authors, the total count of unique words across all branches may exceed 15. For example, we saw that there was a significant overlap in the top words for each author, which is why the total number of unique words exceeded 15. To address this issue, we can add the `scales = "free"` argument to the `facet_wrap()` function. This will allow each facet to have its own scale, ensuring that the top 15 words for each author are displayed correctly:

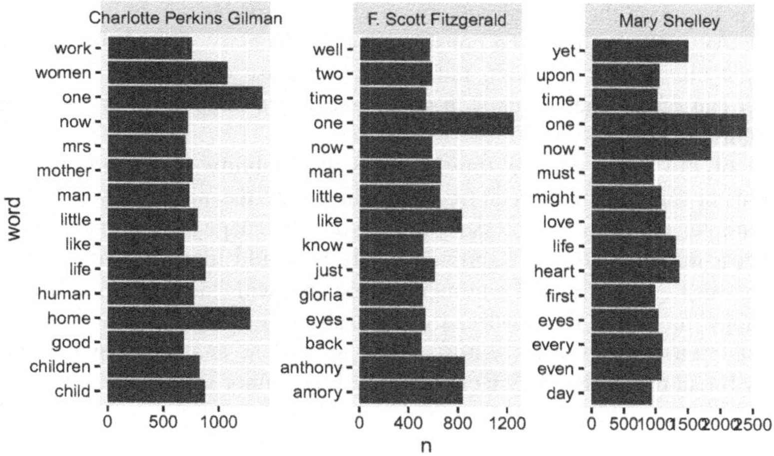

Fig. 5.4 Top 15 words per author, faceted with free scales for accurate comparison

```
# Create a faceted bar plot of the top words for each author with free
↪ scales
author_word_counts |>
  group_by(author) |>
  top_n(15, n) |>
  ggplot(aes(x=word, y=n)) +
  geom_col() +
  coord_flip() +
  facet_wrap(~author,  scales = "free")
```

Figure 5.4 accurately displays the top 15 words for each author, providing a clear comparison of the most frequently used terms in their respective works. We see that each author has a unique set of words that characterize their writing style and thematic focus. While some words may be common across authors, distinctive patterns now emerge.

Having explored the most common words in the texts, we can now delve deeper into the sentiment of the authors' works. Since all of these authors are known for their emotional depth and nuanced character portrayals, it may be interesting to explore the sentiment of their texts. Luckily, once we have placed our text in a tidy format, we can easily apply sentiment analysis techniques.

5.4 Sentiment Analysis

Sentiment analysis operates under the principle of assigning emotional polarity to words, a process that tags individual words within a text as positive, negative, or neutral (Montoyo et al. 2012; Birjali et al. 2021). In a bag-of-words approach, this coding is based on the general perception of the words outside of any surrounding context. As such, it should be considered a rudimentary gauge of the collective emotional undertones in a text, rather than a precise or nuanced understanding. While it can offer general trends in sentiment over a large body of text, there are limitations.

One of the most notable limitations is its inability to accurately interpret negations or sarcasm, which can frequently invert the meaning of a word or phrase. For example, the phrase "not sad" in everyday language might imply a positive or at least neutral state, but sentiment analysis would likely misconstrue this as negative due to the individual sentiments associated with "not" and "sad." Similarly, "not happy" could be misconstrued as neutral, even though in context it conveys dissatisfaction. These issues withstanding, sentiment analysis can still help us identify broad trends that may be indicative of the author's emotional tone.

To undertake sentiment analysis within the "tidytext" framework, we can leverage the `get_sentiments()` function from the "tidytext" package. This function allows us to access a pre-built lexicon of sentiment scores for a wide range of words. With this list, we can then perform a join operation, which merges our main textual data with the sentiment data. This process will match each token from our text to its sentiment equivalent in the `get_sentiments()` tibble, with the result being a new combined dataset that includes sentiment scores alongside the original text data.

There are numerous sentiment lexicons available. For now, we will begin with the "bing" lexicon, which is a simple lexicon that categorizes words as either positive or negative. Let's take a look at this lexicon to see how it is structured:

```
# Get the "bing" sentiment lexicon
get_sentiments("bing")
```

```
# A tibble: 6,786 x 2
   word        sentiment
   <chr>       <chr>
 1 2-faces     negative
 2 abnormal    negative
 3 abolish     negative
 4 abominable  negative
 5 abominably  negative
 6 abominate   negative
 7 abomination negative
 8 abort       negative
 9 aborted     negative
10 aborts      negative
# i 6,776 more rows
```

In our resulting dataset, we see that each word is tagged with a sentiment label, either "positive" or "negative." Note that the "bing" lexicon does not include a neutral category. Instead, words that are not explicitly one of the two categories are left out of the lexicon.

Other lexicons, such as the "afinn" lexicon, provide a more nuanced approach by assigning a numerical score to each word, reflecting the intensity of the sentiment from −5 to 5 (Nielsen 2011). Words with positive connotations are assigned higher scores, while negative words are given lowered ones. Neutral words are assigned a score of zero. For instance, a word like "horrible" may score a -4 for its strong negative connotation, whereas "excellent" might score a 4 or 5 to reflect its strongly positive sentiment. The gradation allows analysts to measure not just the direction of sentiment but also its strength.

The "nrc" lexicon, on the other hand, is more comprehensive, encompassing a broader range of emotions beyond just positive and negative. This lexicon includes categories such as "anger," "anticipation," "disgust," "fear," "joy," "sadness," "surprise," and "trust." One benefit of this lexicon is that it provides a list of emotions that are assigned to each word. For instance, a word like "love" might be tagged with "joy" and "trust," while a word like "hate" might be associated with "anger" and "disgust."

For now, we will stick with the "bing" lexicon but most other lexicons can be used in the same way.

To start using it, we can use a left join to match words in our dataset with those containing sentiments. Since most English words are fairly neutral, we will only have a few words with sentiments. Depending on the type of analysis you want to do, you can include these by doing a `left_join()`, which will mark all the neutral values as `NA`, or use an `inner_join()` to remove them. We are going to use the `left_join()` approach as we do not want to lose any words in our analysis:

```
# Perform a left join to match words in our dataset with those containing
↪  sentiments
books_sentiments <- books_tokens |>
  left_join(get_sentiments("bing"))

# Glimpse the structure of the updated dataset
glimpse(books_sentiments)
```

```
Rows: 745,127
Columns: 4
$ title     <chr> "All the Sad Young Men", "All the Sad Young Men", "All the S~
$ author    <chr> "F. Scott Fitzgerald", "F. Scott Fitzgerald", "F. Scott Fitz~
$ word      <chr> "sad", "young", "men", "_by_", "scott", "fitzgerald", "new",~
$ sentiment <chr> "negative", NA, NA, NA, NA, NA, NA, NA, NA, NA, NA, NA, NA, ~
```

In the resulting dataset, each word has been matched with a sentiment label. Words that are not found in the sentiment lexicon are marked as `NA`.

To get a general overview of how positive or negative the books are overall, we can now simply count the positive and negative values:

```
# Count the number of positive and negative words in the dataset
books_sentiments |>
  count(sentiment)
```

```
# A tibble: 3 x 2
  sentiment      n
  <chr>      <int>
1 negative   62142
2 positive   62629
3 <NA>      620356
```

The resulting table shows an almost equal number of positive and negative words. This balance suggests that the texts may contain a mix of positive and negative sentiments, which is not uncommon in literary works that often explore complex themes and emotions.

Like before, we can also analyze the sentiment of each author's works individually. We can group the data by author and sentiment and then count the number of positive and negative words for each author:

```
# Compute the number of positive and negative words for each author
author_sentiments <- books_sentiments |>
  group_by(author, sentiment) |>
  count()

# Display the sentiment counts for each author
author_sentiments
```

```
# A tibble: 9 x 3
# Groups:   author, sentiment [9]
  author                     sentiment      n
  <chr>                      <chr>      <int>
1 Charlotte Perkins Gilman negative   11896
2 Charlotte Perkins Gilman positive   18301
3 Charlotte Perkins Gilman <NA>      160929
4 F. Scott Fitzgerald      negative   12763
5 F. Scott Fitzgerald      positive   11041
6 F. Scott Fitzgerald      <NA>      139901
7 Mary Shelley             negative   37483
8 Mary Shelley             positive   33287
9 Mary Shelley             <NA>      319526
```

We see that while F. Scott Fitzgerald has an almost equal number of positive and negative words, Mary Shelley has more negative words than positive ones. This difference in sentiment may reflect the darker themes and tragic elements often found in Shelley's works, such as *Frankenstein*. Surprisingly, Charlotte Perkins Gilman has more positive words than negative ones, which may indicate a more optimistic or hopeful tone in her writing than might be expected.

It's noteworthy that the majority of words fall into the neutral category, which highlights the importance of context in understanding the implications of language use. This neutrality suggests that most textual content may serve a more descriptive, rather than emotive, purpose in their writings.

Fig. 5.5 Count of positive and negative sentiment words per author (dodged bars)

To investigate our dataset further, we can segment the texts by each author and visually compare the sentiment distributions. For example, do the works of certain authors consistently skew more negative? Figure 5.5 provides a representation of how the balance between positive and negative sentiments varies across the different authors using a bar graph.

When constructing this graph, it's crucial to adjust the placement of the bars to ensure clarity in our visualization through the `position = 'dodge'` argument in the `geom_col()` function. Without this argument, the bars representing each sentiment would layer on top of one another, making it challenging to differentiate between positive and negative sentiments across authors. We will also filter out any `NA` values to focus solely on the words that have been classified as positive or negative:

```
# Create a colunmn graph to visualize the sentiment distribution for each
↪  author
author_sentiments |>
  na.omit() |>
  ggplot(aes(x=author, y=n, fill=sentiment)) +
  geom_col(position='dodge') +
  coord_flip()
```

To gain more insights into the authors, it would be informative to analyze the relative proportions of positive and negative sentiment within each book, rather than their absolute counts. By comparing these ratios, we can better understand the true sentiment landscape and avoid the distortions that come with varying text sizes.

In the graphical representation, Fig. 5.6 shifts from the `"dodge"` to the `"fill"` option in the `geom-col()` function, changing our bar graph from showing separate, side-by-side bars for each sentiment category to stacked bars that represent the share

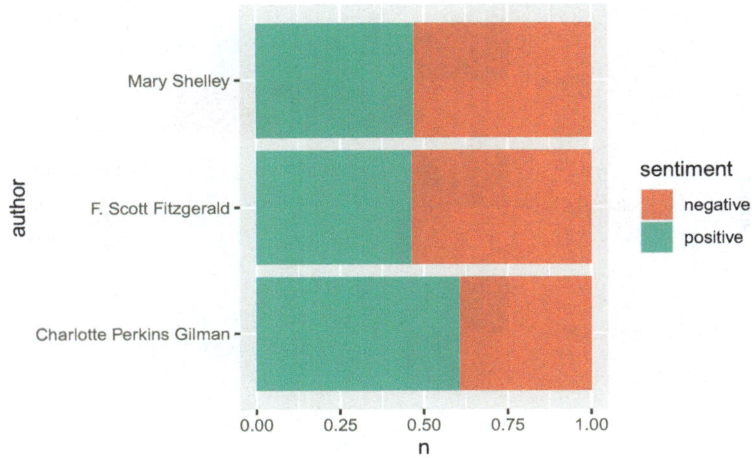

Fig. 5.6 Proportion of positive and negative sentiment words per author (filled bars)

of each sentiment for each work. This method of stacking proportionally adjusts each segment, allowing us to compare the sentiment mix for each work on an even footing. The omission of `NA` values, which represent the neutral words or those not classified by the sentiment dictionary, focuses our attention solely on the words that carry a clear positive or negative charge:

```
# Create a column graph to visualize the sentiment distribution for each
↪ author
author_sentiments |>
  na.omit() |>
  ggplot(aes(x=author, y=n, fill=sentiment)) +
  geom_col(position='fill') +
  coord_flip()
```

Having ratios rather than raw counts allows us to compare the sentiment distribution across authors more effectively. It is ultimately up to you to decide which representation best suits your analytical goals. Both have their advantages. The raw counts provide a more granular view of the sentiment distribution, while the proportions offer a more normalized perspective that accounts for the varying lengths of the texts.

5.5 Part of Speech Tagging

The "bag-of-words" model, while powerful for many kinds of text analysis, operates with a significant blind spot for part of speech tagging: it is agnostic to the grammatical structure and syntactic roles that words play within sentences.

Without this information, interpreting the nuanced dynamics of language becomes challenging. For example, we would not be able to discern whether "dog" is being used as the subject of a sentence, an object of an action, or even as part of an idiomatic expression.

To address this shortcoming, we can enrich our dataset by identifying the parts of speech (POS) for each word. Different branches may make use of adjectives distinctively, perhaps reflecting the unique experiences or offerings at each location. In the "tidytext" package, there is a valuable resource pre-loaded as a tibble named `parts_of_speech`, which encompasses around 205,000 words categorized by the roles they commonly play in sentences (Ward 2006). With such a comprehensive lexicon at our disposal, we can assign POS tags to the words in our dataset using a join, just as we did with sentiment analysis, thus reinstating a layer of syntactic meaning that is otherwise lost in the bag-of-words model.

Two things, however, are worth addressing. First, this dataset only contains the most common usage of a word, which may not represent how it is used in the books. Given the fact that many of the works are from the nineteenth and early twentieth centuries, the language used may differ from contemporary English and may not be accurately represented in the `parts_of_speech` dataset. Newer approaches to parts of speech tagging utilize neural network models that consider the surrounding context of a word, providing a more sophisticated and accurate categorization (Ghosh and Mishra 2020). However, such analysis is beyond the scope of this book. Second, given that the English language allows for words to have multiple uses—where a word can act as a noun in one context and a verb in another—the `parts_of_speech` dataset may contain several entries for a single word. To streamline our analysis and prevent redundancy, we can employ the `distinct()` function, which by default will only select the first occurrence of each word for tagging purposes.

Since we want to keep the values for the column containing the actual part of speech in our join, we will need to pass `.keep_all = TRUE`. Once we have done that, we can add this information to our dataset with a `left_join()`:

```
# Remove duplicate entries in the parts_of_speech dataset while keeping
↪  all values
parts_of_speech_unique <- parts_of_speech |>
  distinct(word, .keep_all=TRUE)

# Perform a left join to match words in our dataset with their parts of
↪  speech
books_pos <- books_tokens |>
  left_join(parts_of_speech_unique, by="word")

# Glimpse the updated dataset
glimpse(books_pos)
```

```
Rows: 745,122
Columns: 4
$ title  <chr> "All the Sad Young Men", "All the Sad Young Men", "All the Sad ~
```

```
$ author <chr> "F. Scott Fitzgerald", "F. Scott Fitzgerald", "F. Scott Fitzger~
$ word   <chr> "sad", "young", "men", "_by_", "scott", "fitzgerald", "new", "y~
$ pos    <chr> "Adjective", "Adjective", "Noun", NA, "Noun", "Noun", "Adjectiv~
```

We are now going to get the top 15 most common nouns for each author. The process for how to visualize these results, shown in Fig. 5.7, should be familiar to you by now. All we need to do is get a count for the author, word, and position. We are also going to use `facet-wrap()` on the author to make it easier to visualize and declare `scales = "free"` to allow both axes to differ for our data:

```
# Get the top 15 most common nouns for each author and create a faceted
↪  column graph
books_pos |>
  filter(pos=="Noun") |>
  count(author, word, pos, sort=TRUE) |>
  group_by(author) |>
  top_n (15) |>
  ggplot(aes(x=reorder(word,n), y=n)) +
    geom_col() +
    coord_flip() +
    facet_wrap(~author, scales="free")
```

We observe that all three authors share certain common terms, suggesting some thematic overlaps in their works. For instance, words like "life" and "man" appear frequently across the texts. These shared nouns suggest that despite their diverse narrative styles and periods of writing, Shelley, Fitzgerald, and Gilman all grapple with fundamental human experiences and social constructs, albeit from different angles and contexts.

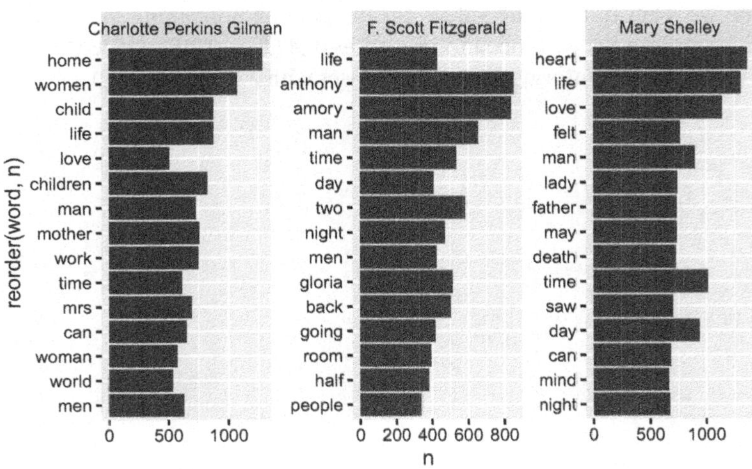

Fig. 5.7 Top 15 most frequent nouns for each author

However, alongside these shared terms are words that reveal the unique focuses and thematic preoccupations of each author. For example, in Charlotte Perkins Gilman's works, "children" emerges as a common noun, underscoring her focus on domestic and familial themes, as well as the societal roles of women and motherhood. This aligns with her well-known exploration of gender dynamics and critiques of the domestic sphere, as seen in pieces like "The Yellow Wallpaper."

Again, an important thing to note is that these nouns are simply the most common. They do not allow us to understand how these work in contexts since we are using a bag-of-words approach. One way to address this is to see which words are unique to each author along with determining when certain words are used together.

5.6 TF-IDF and Pairwise Counts

Counting words in a dataset is a common first step in textual analysis that can provide an initial insight into the most frequently used terms in the given collection of documents. While this approach offers a broad-stroke view of the text, it often reveals a significant amount of overlap among different subsets of the data—such as the titles or authors in our current project.

To address this issue, it becomes necessary to go beyond mere word frequencies and examine the relative importance of words within each corpus. A term could appear frequently across all works and thus might not be particularly illustrative of any single one. For example, the word "love" might be a common word for all authors, but it would be better to uncover terms that are disproportionately represented in one author compared to others, perhaps indicating a unique feature or aspect of that author's writing.

To distill this level of detail from our dataset, we will employ term frequency-inverse document frequency (TF-IDF) analysis (Sparck Jones 1972; Robertson 2004). This requires four steps. First, we need to divide our text into distinct groups. For instance, we could divide our text into groups by author or work. Second, once we have divided the text into groups, we calculate the term frequency. This is simply the count for each word in each group that we have already seen. Third, we will calculate the inverse document frequency. This is a mathematical calculation that provides a "penalty" to words that are common across many or all subgroups. The fourth and final step is a multiplication of these two figures—the term frequency and its inverse document frequency counterpart—to arrive at the tf-idf value for each word in each group. This composite metric encapsulates both the local significance of a term within its subgroup and its importance when viewed within the broader context of the entire corpus. Higher tf-idf values signal terms that are potentially more descriptive and indicative of the unique characteristics of their respective subgroups, and lower values correspond to terms with less distinguishing power.

To implement this process, we will use the `bind_tf_idf()` function from the "tidytext" package. This function will automatically calculate the term frequency-inverse document frequency for each word in our dataset. This function takes three arguments with the pipe operator: the column containing each token, the group we want to divide by (in our case, the author), and the text data itself. We will store the result in a new tibble called `books_tfidf`. Remember that when doing `count()`, this information is stored in the "n" column:

```
# Calculate the term frequency-inverse document frequency for each word
↪  in the dataset
books_tfidf <- books_tokens |>
  count(author, word) |>
  bind_tf_idf(word, author, n)

# Glimpse the structure of the updated dataset
glimpse(books_tfidf)
```

```
Rows: 60,039
Columns: 6
$ author <chr> "Charlotte Perkins Gilman", "Charlotte Perkins Gilman", "Charlo~
$ word   <chr> "_absolutely", "_academy", "_ached_", "_adobe_", "_alive_", "_a~
$ n      <int> 1, 1, 1, 1, 1, 8, 1, 3, 3, 4, 7, 1, 2, 3, 1, 1, 10, 1, 6, 1, 1,~
$ tf     <dbl> 5.232178e-06, 5.232178e-06, 5.232178e-06, 5.232178e-06, 5.23217~
$ idf    <dbl> 1.0986123, 1.0986123, 1.0986123, 1.0986123, 1.0986123, 0.405465~
$ tf_idf <dbl> 5.748135e-06, 5.748135e-06, 5.748135e-06, 5.748135e-06, 5.74813~
```

You will see that our function has calculated the term frequency, inverse document frequency, and term frequency-inverse document frequency. Most of the values are not very high. As a result, we will arrange them in descending order before graphing them and only graph the top ten words per author, as shown in Fig. 5.8:

```
# Create a faceted column graph of the top ten words for each author
↪  based on tf-idf
books_tfidf |>
  group_by(author) |>
  arrange(desc(tf_idf)) |>
  top_n(10) |>
  ggplot(aes(x=reorder(word, n), y=n)) +
    geom_col() +
    coord_flip() +
    facet_wrap(~author, scales="free")
```

We now see that the majority of the top ten words by tf-idf are names of characters. The frequency of these names within a single work (high term frequency) combined with their rarity across different works by other authors (low document frequency) results in a high tf-idf score. As a result, it is not surprising as character names are often unique to each author and work.

Along with the tf-idf, we may also be interested in what words show up together for a given group of texts. For instance, let's say we are interested in finding out which words are most likely to appear together in the works of F. Scott Fitzgerald.

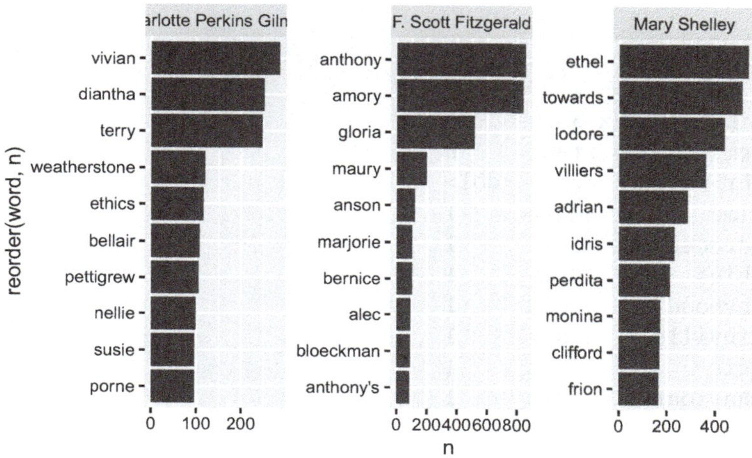

Fig. 5.8 Top ten words with the highest TF-IDF scores for each author

To do this, we can use the `pairwise_count()` function in the "widyr" package. If you have not already done so, make sure to install and load the "widyr" package before running the code below:

```
# install.packages("widyr")
library(widyr)
```

The `pairwise_count()` function takes two main arguments with the pipe operator. The first is the column containing our words. The second is the grouping variable, which in this case is the author. We will store the results in a new tibble called `books_pairwise`. An optional yet often helpful addition to our function call is the `sort = TRUE` argument. Enabling this feature organizes our output in descending order, ensuring that the most frequently co-occurring words rise to the top of our results. This sorting lays bare the strongest word partnerships, making it immediately apparent which terms are most commonly linked in the discourse encapsulated by our texts.

Calculating the correlation for all words can be computationally intensive. For now, we will only look at words that have a tf-idf value greater than 0.1. This will allow us to focus on the most important words in our dataset:

```
# Calculate the pairwise count for words that appear more than 10000
↪  times in the dataset
author_pairwise <- books_tfidf |>
  filter(tf_idf > 0.0001) |>
  pairwise_count(word, author, sort=TRUE)
```

```
# Get top 10 words test that appear together
head(author_pairwise, 10)
```

```
# A tibble: 10 x 3
   item1        item2      n
   <chr>        <chr> <dbl>
 1 alima        adult     1
 2 androcentric adult     1
 3 babies       adult     1
 4 babyhood     adult     1
 5 bainville    adult     1
 6 basic        adult     1
 7 behaviour    adult     1
 8 bellair      adult     1
 9 can't        adult     1
10 celis        adult     1
```

5.7 Understanding Topic Modeling

When faced with the daunting task of analyzing and understanding massive volumes of textual data, manual reading becomes infeasible due to the sheer scale and time constraints involved. Topic modeling is a computational solution to systematically categorize and make sense of this data. Among the various algorithms designed for this purpose, latent Dirichlet allocation (LDA) stands out as one of the most widely implemented and effective methods (Jelodar et al. 2019).

LDA is a probabilistic model that identifies clusters of words that frequently occur together across the corpus and organizes these clusters into what we refer to as "topics." To illustrate, within a compendium of sports-related articles, words such as "baseball," "pitcher," "homerun," and "inning" might frequently co-occur. Conversely, another cluster might include terms like "tackle," "touchdown," and "quarterback," suggesting a different topic centered around football. Once the LDA model is established, it calculates the mixture of topics for each document, which is presented as a distribution of topics within that text. In practice, most texts exhibit a mix with varying proportions from multiple topics, reflecting the multifaceted subject matter within the document.

It's important to note that LDA assigns numerical labels to these topics (such as Topic 1 and Topic 2) and does not inherently understand or assign semantic meaning to them. The task of interpreting these topics and giving them descriptive labels rests with the researcher or analyst. The selection of the number of topics ("k") is also a critical decision that can significantly influence the granularity and utility of the model. Choosing too few topics may oversimplify the model, failing to capture the nuanced themes present in the data. Conversely, setting the number too high can result in fragmented and overly specific topics that overlap or do not meaningfully contribute to the understanding of the corpus. The optimal number often requires iterative testing and a balance between detail and interpretability.

Although the tidy text format is wonderful for doing many forms of analyses, sometimes working with other packages is easier. For instance, in topic modeling, it is easier to work with libraries specifically geared toward the method. However, these often require that your data be in a different structure. Fortunately, the "tidytext" package allows you to convert between formats that other packages may need to use. For topic modeling, we will be using the "topicmodels" package, which requires our data to be placed in a document term matrix (Grün et al. 2023). Let's install it and load it:

```
# install.packages("topicmodels")
library(topicmodels)
```

To work with the "topicmodels" package, we need to convert our data into a document term matrix (DTM). A DTM is one of the most common formats for structuring text. In a DTM, each column represents the words within the document, with the exception of the first column, which represents the document ID. The values underneath each column represent the frequency of the given word in that document. To turn our tidy data frame into a DTM, we can use the `cast_dtm()` function. It will look for a column that contains the grouping variable, a list of words, and a column containing the counts for the words in that grouping variable. As before, we can use `count()` to get this information.

We are going to look at the topics for each of the authors. So, let's see this in practice by doing our count and casting our results to a DTM:

```
# Count the occurrences of each word within each author and cast the
↪  results to a DTM
authors_dtm <- books_tokens |>
  count(author, word) |>
  cast_dtm(author, word, n)
```

Now that our data is in a DTM, we need to call the `LDA()` function—remember that LDA is just the most common form of topic modeling—from the "topicmodels" package. Before we do that though, there are two important considerations to understand.

First, LDA involves random initialization of its algorithms, which means you might get slightly different results each time you run the model. To ensure reproducible results, it's crucial to set a random seed before running LDA using `set.seed(your_preferred_number)`. This ensures the random number generator produces the same sequence, resulting in consistent results across runs.

Second, choosing the number of topics (k) is a critical decision that significantly impacts the model's usefulness. While we will start with k=5, this choice requires careful consideration. Too few topics can oversimplify complex textual relationships and group unrelated terms together, while too many topics may create artificial distinctions and make interpretation difficult.

Please note that every time you run an LDA model, you might get slightly different results because the initialization of the algorithms involves some randomness. Thus, to get reproducible results, you might want to set a seed before running your LDA command using `set.seed(your_preferred_number)`. This ensures that the random number generator gives the same sequence of numbers, resulting in the same initial conditions and, therefore, the same results in our model:

```
# Set a seed to ensure reproducibility of results
set.seed(123)

# Fit an LDA model to the document-term matrix with five topics
# Create an LDA topic model with 5 topics
author_topics <- authors_dtm |>
  LDA(k = 5)
```

We now have a topic model with five topics. However, this doesn't tell us much. The process of interpreting the results from an LDA topic model can be facilitated by transitioning the model output into a more human-readable format, an operation easily handled by the `tidy()` function. This function, part of the "broom" package, adeptly converts statistical analysis objects into tidy data frames. When dealing with LDA topic models, we can use the `tidy()` function from the "broom" package to extract two primary matrices: "beta" and "gamma," which are crucial for understanding the composition of topics and their prevalence within documents.

Let's go ahead and load the `broom` package, which provides the `tidy()` function for extracting LDA matrices:

```
# install.packages("broom") # Install if you don't have it
library(broom)
```

Now that we have loaded `broom`, we can use the `tidy()` function to extract the LDA matrices. We will begin by looking at the "beta" matrix, which represents the probability per-topic-per-word probability. High "beta" values for words within a certain topic suggest that these words are particularly influential in shaping the thematic outline of what could be inferred as a baseball-related topic. Typically, the results are sorted in descending order of "beta" to pinpoint the most defining words for each topic. The words with the highest "beta" values for a topic can be seen as that topic's "keywords," which are the most characteristic or frequent within that specific cluster:

```
# Extract the beta matrix from the LDA model and arrange it in descending
↪  order
author_topics |>
  tidy(matrix = "beta") |>
  arrange(desc(beta))
```

```
# A tibble: 182,005 x 3
   topic term       beta
   <int> <chr>     <dbl>
 1     5 one      0.0105
 2     4 home     0.0101
 3     4 one      0.00847
 4     1 children 0.00814
 5     3 one      0.00776
 6     1 women    0.00738
 7     5 now      0.00738
 8     1 man      0.00720
 9     1 little   0.00661
10     4 human    0.00646
# i 181,995 more rows
```

We can see that for each topic, we now have a beta value showing how much each term contributes to a topic. We can look at the top values for any given topic by filtering and arranging. Let's see the top ten words for topic 4:

```
# Extract the beta matrix and filter for topic 4
author_topics |>
  tidy(matrix = "beta") |>
  filter(topic == 4) |>
  arrange(desc(beta)) |>
  head(10)
```

```
# A tibble: 10 x 3
   topic term      beta
   <int> <chr>    <dbl>
 1     4 home    0.0101
 2     4 one     0.00847
 3     4 human   0.00646
 4     4 child   0.00621
 5     4 well    0.00523
 6     4 new     0.00511
 7     4 mother  0.00484
 8     4 love    0.00471
 9     4 world   0.00467
10     4 life    0.00448
```

We see that all of these words are related to the topic of family and domestic life. Figure 5.9 plots the top ten words for each topic using this same approach:

```
# Extract the beta matrix and plot the top ten words for each topic
author_topics |>
  tidy(matrix="beta") |>
```

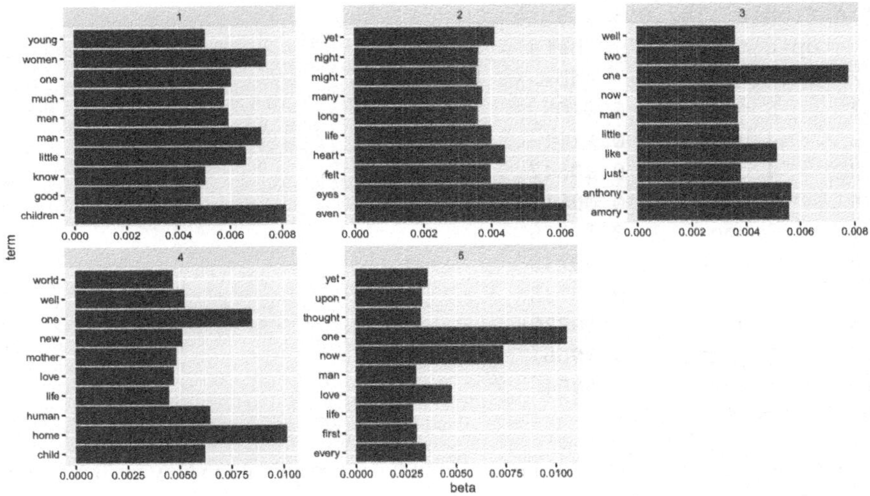

Fig. 5.9 Top ten words contributing most to each identified topic (beta values)

```
  group_by(topic) |>
  top_n(10) |>
  ggplot(aes(x=term, y=beta)) +
  geom_col() +
  coord_flip() +
  facet_wrap(~topic, scales="free")
```

Having established a clearer understanding of the constituent words and their respective probabilities within each topic through the "beta" matrix, we can now turn our attention to the "gamma" matrix to investigate the role of topics within our collection of texts. The "gamma" values are of particular interest as they provide a numerical depiction of the makeup of topics within individual documents—in our scenario, the authors' corpus. For instance, an author's corpus may be composed of a certain percentage of Topic 1, which perhaps talks about love and relationships, and another percentage of Topic 2, which might focus on nature and the environment. By examining the "gamma" matrix, we can discern the distribution of topics within each document, shedding light on the thematic composition of the texts.

The only change we need to make to get the gamma matrix is to our `matrix` argument. We will also arrange the data in descending order based on the gamma values to see which topics are most prevalent for each author:

```
# Extract the "gamma" matrix from the topic model results and arrange it
↪  in descending order
author_topics |>
  tidy(matrix="gamma") |>
  arrange(desc(gamma))
```

```
# A tibble: 15 x 3
   document                  topic       gamma
   <chr>                     <int>       <dbl>
 1 F. Scott Fitzgerald           3 0.930
 2 Mary Shelley                  5 0.561
 3 Charlotte Perkins Gilman      4 0.503
 4 Charlotte Perkins Gilman      1 0.497
 5 Mary Shelley                  2 0.439
 6 F. Scott Fitzgerald           1 0.0702
 7 Mary Shelley                  4 0.000176
 8 Mary Shelley                  1 0.0000994
 9 Charlotte Perkins Gilman      2 0.0000150
10 Mary Shelley                  3 0.00000623
11 F. Scott Fitzgerald           4 0.00000607
12 F. Scott Fitzgerald           2 0.00000534
13 Charlotte Perkins Gilman      3 0.00000285
14 F. Scott Fitzgerald           5 0.000000570
15 Charlotte Perkins Gilman      5 0.000000489
```

It looks like topic 4 is the most prevalent for Charlotte Perkins Gilman. In contrast, the top topic for F. Scott Fitzgerald is 5, and for Mary Shelley, it is 1. Figure 5.10 visualizes this by creating a bar chart that shows the gamma values for each topic in each document. This will allow us to see the distribution of topics within each author's works. Note that, by default, each collection of works

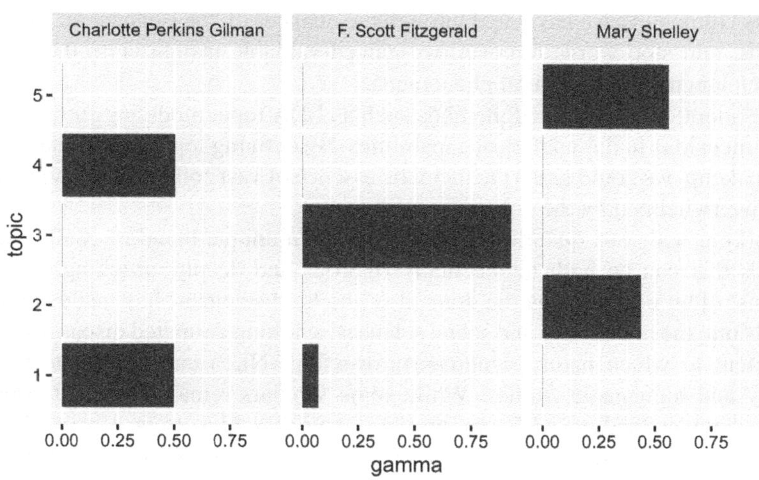

Fig. 5.10 Distribution of topics (gamma values) within the works of each author

is considered a document in the LDA model. As a result, we need to rename the document column to author to make it easier to understand:

```
# Create a faceted column graph of the gamma values for each author
author_topics |>
  tidy(matrix="gamma") |>
  rename(author = document) |>
  group_by(topic) |>
  arrange(desc(gamma)) |>
  ggplot(aes(x=topic, y=gamma)) +
    geom_col() +
    coord_flip() +
    facet_wrap(~author)
```

5.8 Conclusion

In this chapter, we have embarked on a journey through the realms of text analysis, exploring the multifaceted nature of text as data within the field of cultural analytics. Employing the tidy text framework, we have seen the transformative process of converting text into a structured, analyzable form, allowing us to shift from qualitative reading to quantitative analysis. Whether dealing with the tokenization of words, the distillation of sentiment, or the extraction of metadata characteristics, each step has brought us closer to understanding the rich patterns embedded within the works of authors like Mary Shelley, F. Scott Fitzgerald, and Charlotte Perkins Gilman. As we navigated through the nuances of "bag-of-words," sentiment analysis, and stop words removal, we gained valuable insights into the interplay between language and cultural phenomena.

The adoption of advanced methods such as LDA topic modeling further opened new dimensions in our analytical capabilities. By defining topics and understanding their makeup, we could grasp the thematic essence of vast collections of text without being overwhelmed by the volume.

Working with vast quantities of text presents a unique blend of challenges and rewards that can be both intellectually rigorous and deeply enriching. Statistical analysis, often perceived as the domain of the hard sciences, has made significant inroads into the humanities and social sciences, sparking animated discussions about the extent to which natural language processing (NLP) can illuminate complex literary and humanistic studies. While some scholars remain skeptical about the depth of understanding NLP can provide, it undeniably offers a powerful point of departure for sophisticated textual analysis in an era increasingly dominated by digital data.

Chapter 6
Understanding Regression

Abstract The chapter presents an accessible exploration of regression analysis tailored to the cultural research domain. The treatment begins with an overview of linear regression, elucidating the method's core objectives, assumptions, and its practical utility for inferential statistics and prediction. By leveraging a popular dataset, the chapter illustrates the process of building, interpreting, and validating linear regression models, emphasizing their applicability in drawing robust cultural inferences. The narrative progresses to logistic regression, a technique adept at modeling binary outcomes. Throughout, the chapter balances statistical rigor with practical guidance, equipping readers with the tools to both understand and apply regression analyses to cultural datasets. The chapter details insights into the design intentions of the Pokémon series of games, as inferred through statistical patterns, offering a compelling example of how quantitative methods can yield qualitative understandings in cultural studies.

Keywords Regression analysis · Linear regression · Logistic regression · Predictive modeling · Cultural research · Data analysis · Statistical assumptions · Model validation · Quantitative methods

6.1 Introduction to Regression Analysis

The term "regression," especially within the context of statistics and data analysis, might invoke a sense of intimidation for those whose expertise lies in the humanities and qualitative social sciences. However, this perception of complexity often dissipates with a clearer understanding of the concept. At its essence, regression analysis is a statistical method that can be boiled down to two fundamental objectives: first, to analyze and infer the collective behavior of a group, or population, by examining a representative sample; and second, by studying this sample, we can make educated guesses about the wider group's characteristics and tendencies.

The second objective is rooted in the ability to make predictions. Through regression, we can estimate the value of a dependent variable—often called the outcome variable—based on the known values of one or more independent variables, which

N. Siddiqui, *Cultural Analytics in R: A Tidy Approach*, Use R!,
https://doi.org/10.1007/978-3-031-96618-7_6

are commonly referred to as predictors. For example, we might use regression to predict a student's final grade based on their hours of study, attendance, and participation scores.

In the realm of cultural research, the advent of digitization and datafication has led to a wealth of information ripe for this type of analysis. Regression methods, by facilitating the examination of relationships and trends within this burgeoning data, are thus poised to become even more instrumental in the field. Despite this, there is a tendency for many cultural analytic scholars to leapfrog over these foundational methods in favor of more complex machine learning techniques.

This chapter aims to demystify regression and demonstrate its applicability and power within the study of culture. You will delve into the intricacies of linear regression—a method that explores the relationship between variables to predict continuous outcomes, such as the number of book sales based on past trends and marketing efforts. Then, you'll venture into the territory of logistic regression, which is used for predicting binary outcomes, such as whether a film will be a hit or a miss based on viewer ratings and budget.

Although linear regression is a powerful technique, it comes with a myriad of caveats and assumptions about what data should look like that researchers often overlook, misunderstand, and/or fail to report. Furthermore, like other models, linear regression merely provides an estimate for observed phenomena. When presenting our information to a broader audience, we need to make clear just how well our data actually "fits" the model. Thus, along with overviewing linear regression, this chapter will also provide guidance for checking model assumptions and reporting them.

By the conclusion of this chapter, your comprehension of how to utilize regression models in the sphere of cultural data should be robust and well-grounded. The focus will not be on the intricate mathematical underpinnings but rather on the practical application and interpretation of regression analysis to ensure you can draw reliable and meaningful inferences.

One thing to acknowledge is that this chapter will diverge somewhat from the tidyverse-focused approach that we have emphasized so far. Tidy data is a powerful technique for organizing and presenting data in a way that simplifies analysis, and it benefits from a large amount of community-developed resources. However, while some tools—such as "broom"—make it easier to work with regression models in the tidyverse ecosystem, in practice, you will find that the statistical literature on regression has its own unique notation and terminology that may not always align perfectly with the principles of tidy data. These conventions are deeply rooted in the history of statistical thought and practice.

Understanding regression within this broader context is vital for several reasons: First, it will enable you to read and comprehend the vast array of existing literature and research that employs regression analysis; second, it will equip you with a versatile skill set that is not confined to a single set of tools, enhancing your ability to communicate with statisticians and other data scientists who might not use the tidyverse; and lastly, by bridging the gap between the tidyverse and traditional statistical methods, you will be better prepared to tackle complex questions with regression models, whether they arise from historical datasets or contemporary cultural inquiries. This balanced perspective is crucial for effectively applying regression analysis to the multifaceted challenges and datasets that embody the modern study of culture.

6.2 Loading Pokemon Battle Data

The data we will look at in this chapter comes from a popular dataset on Kaggle that consists of Pokémon characteristics alongside simulated battle results (Challenge, n.d.). If you are not familiar with the Pokémon games, don't worry; their overall design is relatively simple. Throughout the series, players take on the role of a Pokémon trainer that tries to "capture" Pokémon and employ them to battle other creatures.

Our dataset itself consists of three files. The first file contains data about Pokémon and their attributes derived from the actual games, such as defense, attack, speed, and hit points (health). These statistics are crucial as they determine the strengths and weaknesses of each Pokémon. The second file has a list of simulated battles between two Pokémon and the winner of each battle. It's important to note that while the Pokémon characteristics are from the game, this battle data was generated by a custom algorithm and simplified, meaning it doesn't take into account all actual game characteristics. This file provides a synthetic historical record from which we can learn the relationship between the Pokémon's attributes and their performance in these simulated battles. Finally, we have a file with battle pairings but no winner, presenting an opportunity to predict outcomes based on the patterns learned from the simulated data.

The goal for analyzing this data will be twofold: First, we will try to make sense of the broader culture of gaming by determining if and how the designers of the Pokémon games sought to appeal to new players. Role-playing games, such as those in the Pokémon series, derive from pen-and-paper games like Dungeons and Dragons. These games often have certain "classes" of characters—or in this case Pokémon—with statistics geared toward different roles or playing styles. For instance, in most role-playing games, one character might have large defensive stats to help take damage but have little attack power, while another might be used as a "support" character.

When players are new to role-playing games, they often find determining these roles difficult to grasp. If we suspect that the Pokémon game designers wanted to appeal to a broader audience, we would expect that these conventions would be subdued. For instance, instead of a Pokémon that has high defense and low attack, we might see that certain Pokémon simply have high statistics across the board. This would make it easier for players to categorize Pokémon in simpler terms, such as strong or weak, rather than through more complex conventional role-playing roles.

Second, we will seek to create a model that will predict which Pokémon will win battles in the future. Such a model is not just an academic exercise; it mirrors the kinds of predictive models used in various industries, from finance to health care, to forecast outcomes based on historical data.

Let's get started by loading our data with the `read_csv()` function and viewing it using `glimpse()`. The dataset codes missing data as a blank value. As a result, we need to convert these to R's NA type by passing a blank to the `read_csv()` function's na argument.

```
# Import Pokemon dataset converting blank values to NA
pokemon <- read_csv("data/ch5_pokemon.csv", na = "")

# Examine dataset structure showing columns and data types
glimpse(pokemon)
```

```
Rows: 800
Columns: 12
$ id         <dbl> 1, 2, 3, 4, 5, 6, 7, 8, 9, 10, 11, 12, 13, 14, 15, 16, 17, ~
$ name       <chr> "Bulbasaur", "Ivysaur", "Venusaur", "Mega Venusaur", "Charm~
$ type_1     <chr> "Grass", "Grass", "Grass", "Grass", "Fire", "Fire", "Fire",~
$ type_2     <chr> "Poison", "Poison", "Poison", "Poison", NA, NA, "Flying", "~
$ hp         <dbl> 45, 60, 80, 80, 39, 58, 78, 78, 78, 44, 59, 79, 79, 45, 50,~
$ attack     <dbl> 49, 62, 82, 100, 52, 64, 84, 130, 104, 48, 63, 83, 103, 30,~
$ defense    <dbl> 49, 63, 83, 123, 43, 58, 78, 111, 78, 65, 80, 100, 120, 35,~
$ sp_attack  <dbl> 65, 80, 100, 122, 60, 80, 109, 130, 159, 50, 65, 85, 135, 2~
$ sp_def     <dbl> 65, 80, 100, 120, 50, 65, 85, 85, 115, 64, 80, 105, 115, 20~
$ speed      <dbl> 45, 60, 80, 80, 65, 80, 100, 100, 100, 43, 58, 78, 78, 45, ~
$ generation <dbl> 1, 1, 1, 1, 1, 1, 1, 1, 1, 1, 1, 1, 1, 1, 1, 1, 1, 1,~
$ legendary  <lgl> FALSE, FALSE, FALSE, FALSE, FALSE, FALSE, FALSE, FALSE, FAL~
```

As we observe in this table, we have the following columns:

- **id**: A generic id for the Pokémon
- **name**: The name of the Pokémon
- **type_1**: Each Pokémon has at least one type representing the kind of Pokémon it is (Water, Bug, Fire, etc.). This is called a primary type.
- **type_2**: Some Pokémon have more than one type. This is known as a secondary type.
- **hp**: HP is short for "hit points" and reflect the overall health of each Pokémon.
- **attack**: A value representing how powerful a Pokémon is when doing physical attacks.
- **defense**: A value representing the defensive capabilities of Pokémon when facing physical attacks.
- **sp_attack**: Pokémon have special moves that they can use for attacks. This is a representation of the power of those attacks.
- **sp_defense**: When Pokémon face special attacks, they use their special defense characteristic. This is a representation of that defense.
- **speed**: Value representing how much speed each Pokémon has.
- **generation**: The introduction of a new set of Pokémon corresponds with a new generation. Although these are not official designations, they are often used by the fan community.

We need to do a little bit of cleanup of our data before exploring it. The "generation" column is a category, but the `read_csv()` function loads this as a numeric value. To fix this, we will convert this column to a factor.

```
# Transform generation column to categorical factor type
pokemon <- pokemon |>
  mutate(generation = as.factor(generation))
```

We now have a workable dataset. We can now look into how understanding the statistical issues of different Pokémon can help us explore game design, culture, and marketing.

6.3 Exploring Pokemon Data

Before we dive into the nuances of our theory regarding the complexity of the Pokémon games—specifically, the hypothesis that game designers may have intentionally simplified certain elements to make the series more accessible to newcomers to the role-playing game (RPG) genre—we must first familiarize ourselves with the nature of the data at our disposal.

To begin dissecting our theory, we will start by analyzing the most basic categorization within the game: the types of Pokémon. Each Pokémon is assigned a primary type that reflects its elemental nature, such as Fire or Water, which affects its combat abilities, strengths, and weaknesses. Some Pokémon also possess a secondary type, adding an additional layer of strategic depth to their use in battle. Let's create a bar graph showing the first type.

```
# Visualize frequency distribution of primary Pokemon types
pokemon |>
  ggplot(aes(x = type_1)) +
  geom_bar() +
  coord_flip()
```

Figure 6.1 shows that types such as Water, Normal, Bug, and Grass are particularly prevalent. This pattern could suggest various design intentions: Perhaps these types are meant to be more accessible for early encounters, or maybe they are distributed generously to ensure a balanced experience for new players.

Along with the primary type, some Pokémon also have a secondary type. Certain Pokémon types may be paired more frequently with others to balance strengths and weaknesses, create particular thematic combinations, or enhance the strategic depth of battles. By constructing a bar graph of the most common secondary types, we can gain insight into the pairing strategies used by the game designers. For example, we might discover that Flying is a frequent secondary type due to its combination with various other types to create creatures that resemble creatures of myth or common wildlife.

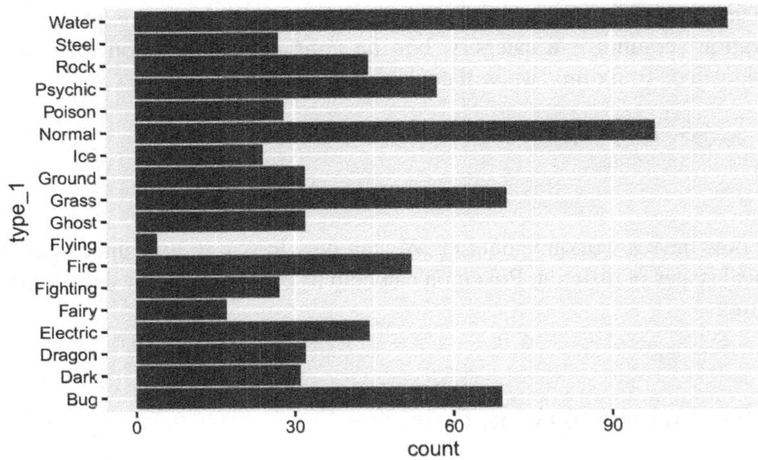

Fig. 6.1 Frequency distribution of primary Pokémon types

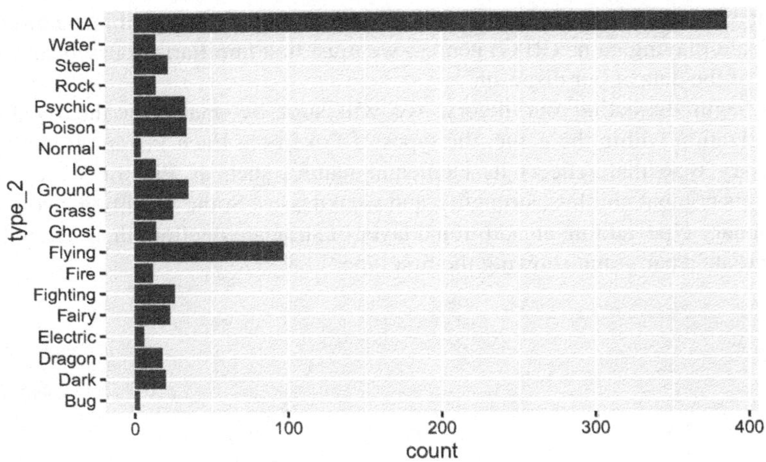

Fig. 6.2 Frequency distribution of secondary Pokémon types (including NA for single-typed Pokémon)

```
# Visualize frequency distribution of secondary Pokemon types
pokemon |>
  ggplot(aes(x = type_2)) +
  geom_bar() +
  coord_flip()
```

As seen in Fig. 6.2, a distinct pattern emerges—the Flying type stands out as the most frequently occurring secondary type. Additionally, the large number of

Pokémon without a secondary type, as indicated by the significant count of NA (Not Available) in the data, suggests that a single-typed Pokémon is not uncommon and may represent a design choice to maintain simplicity and accessibility for new players.

Moving our focus to a particular subset within the Pokémon universe, we find the category of "legendary" Pokémon. These Pokémon are typically characterized by higher stats, unique abilities, and limited availability. One question worth exploring is whether there is a correlation between specific Pokémon types and the likelihood of them being classified as legendary. Are certain types more esteemed or imbued with a sense of rarity and power within the game's lore?

To explore this, we can utilize the `position` argument through the `geom_bar()` function. This will allow us to not just count occurrences but to understand proportions and make comparisons across types on an equal footing.

```
# Analyze legendary Pokemon distribution across primary types
pokemon |>
  ggplot(aes(x = type_1, fill = legendary)) +
  geom_bar(position = "fill") +
  coord_flip()
```

Figure 6.3 reveals a compelling trend: Pokémon with the Flying and Dragon types are disproportionately represented in the legendary category. This observation aligns with the lore of the Pokémon series, where legendary Pokémon are not only rare but often possess attributes and powers that elevate them above the average creatures inhabiting the game world. However, their numerical scarcity prompts the question: What tangible advantages might legendary Pokémon provide to players who capture them? It's reasonable to hypothesize that legendary Pokémon,

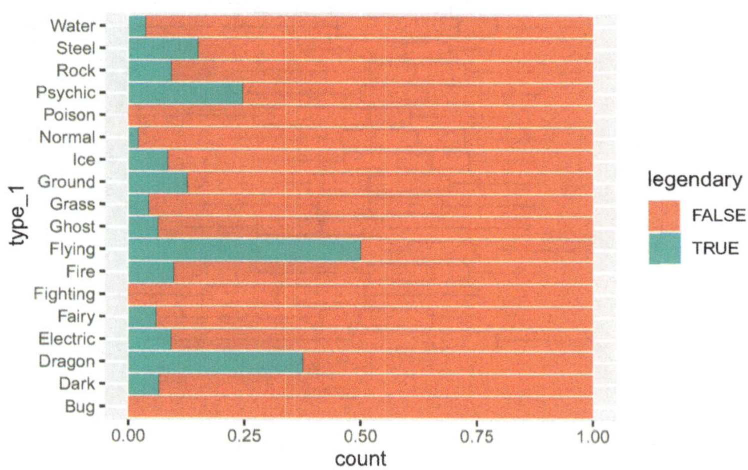

Fig. 6.3 Proportion of legendary Pokémon within each primary type

including those of the Flying and Dragon types, are endowed with superior strength, potentially evidenced by higher attack stats.

To investigate this hypothesis, we can create a box plot. Should the median attack value for Flying and Dragon types be notably higher than that of other types, it would suggest that the game developers have intentionally crafted these Pokémon to be more formidable thereby justifying their legendary status. Such a finding would not only validate players' perception of these Pokémon as valuable assets in their quest to become champion trainers but would also offer insights into how the game balances challenge and reward.

```
# Compare attack stat distributions across Pokemon types
pokemon |>
  ggplot(aes(x = type_1, y = attack)) +
  geom_boxplot() +
  coord_flip()
```

Figure 6.4 shows that the majority of these creatures have attack stats that fall within the moderate range of 50 to 75. Such a distribution ensures a level playing field for the majority of battles, allowing players with diverse Pokémon teams to compete on relatively equal footing. In stark contrast to this standard range are the Dragon-type Pokémon, whose attack stats not only exceed the average but achieve the highest median among all types.

Does the advantage that Dragon Pokémon have only impact their offense or do they also have greater health? We can do a similar box plot to look at how type and HP intersect.

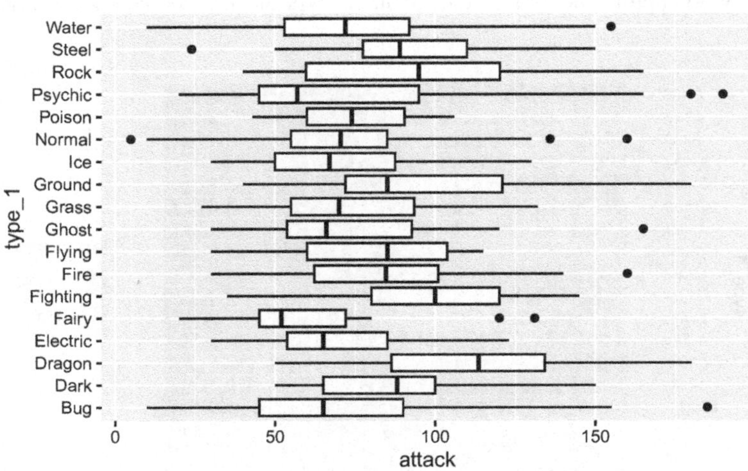

Fig. 6.4 Distribution of attack stats across primary Pokémon types

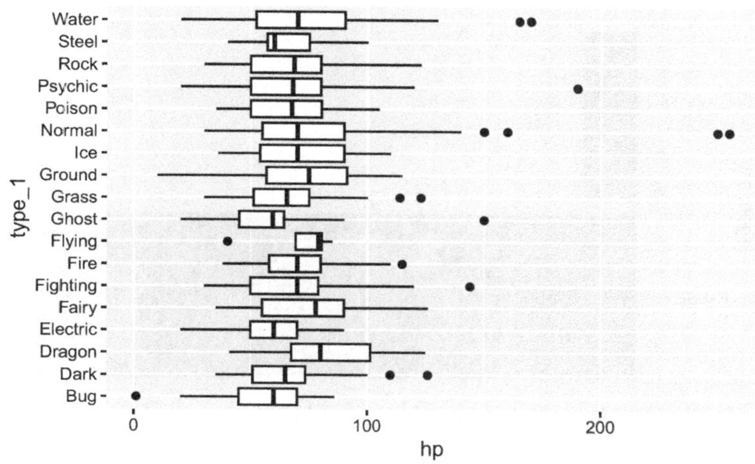

Fig. 6.5 Distribution of HP (Hit Points) across primary Pokémon types

```
# Compare HP stat distributions across Pokemon types
pokemon |>
  ggplot(aes(x = type_1, y = hp)) +
  geom_boxplot() +
  coord_flip()
```

The data shown in Fig. 6.5 suggests that Dragon-type Pokémon are uniquely advantaged not only in terms of their formidable attack capabilities but also with respect to their hit points (HP), which is indicative of their overall health and survivability in battle. This dual advantage positions them as powerhouses capable of taking significant damage while also dishing out heavy hits in return.

This pattern of high attack and HP aligns with the theory that the Pokémon series has been crafted to appeal to a wide range of players, including those who may not be deeply versed in the strategic intricacies of role-playing games (RPGs). In many traditional RPGs, character classes are often designed with specialized roles in mind: Tanks absorb damage; healers provide support; damage dealers focus on offense. However, the relative all-around strength of Dragon-type Pokémon suggests a deviation from this convention. Instead of pigeonholing creatures into rigid roles, the designers appear to have created certain Pokémon, like those of the Dragon type, to be versatile and straightforwardly powerful.

Let's do one more exploratory analysis to see how many legendary Pokémon we have in total. We already saw that Dragon Pokémon are more likely to be legendary, but we should probably determine just how rare legendary Pokémon are overall. We can use a simple bar graph to do a count of the "legendary" column.

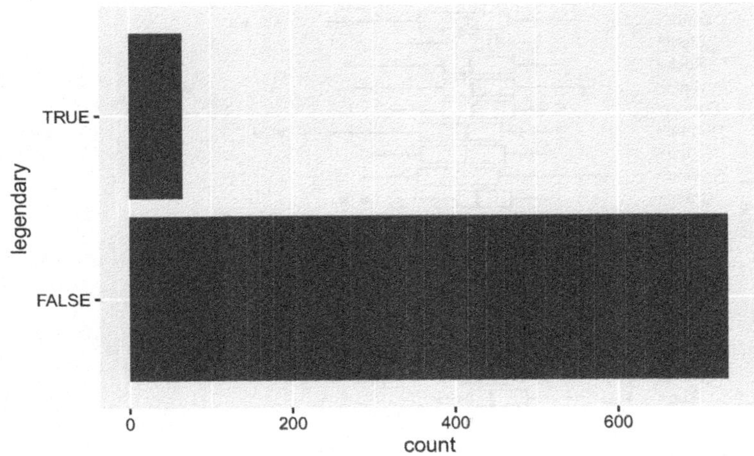

Fig. 6.6 Count of legendary vs. non-legendary Pokémon

```
# Analyze frequency distribution of legendary vs non-legendary Pokemon
pokemon |>
    ggplot(aes(x = legendary)) +
    geom_bar() +
    coord_flip()
```

Figure 6.6 shows that legendary Pokémon are not only strong in terms of their statistics but that they are also rare. However, to fully understand the role of Pokémon within the wider context of the game and its potential departure from traditional RPG archetypes, a deeper analysis is required. We must examine the relationships and interactions between different statistical characteristics across all Pokémon, not just the legendary ones. By analyzing how various stats—such as defense, special attack, and special defense—correlate with each other, and with the Pokémon types, we can discern whether the game leans toward a system of specialized roles or if it favors versatility. For example, if we observe that Pokémon with high attack stats also tend to have high defense stats, it could suggest that the game eschews traditional class-based specialization in favor of creating powerful all-rounders.

6.4 Graphing Linear Regression

Attack seems to be an interesting property for each Pokémon, since it is what is used to inflict damage to other creatures. In conventional role-playing games, attack tends to increase with speed. In other words, there are some creatures that are slow and defensive, while others are quick and offensive. If we were interested

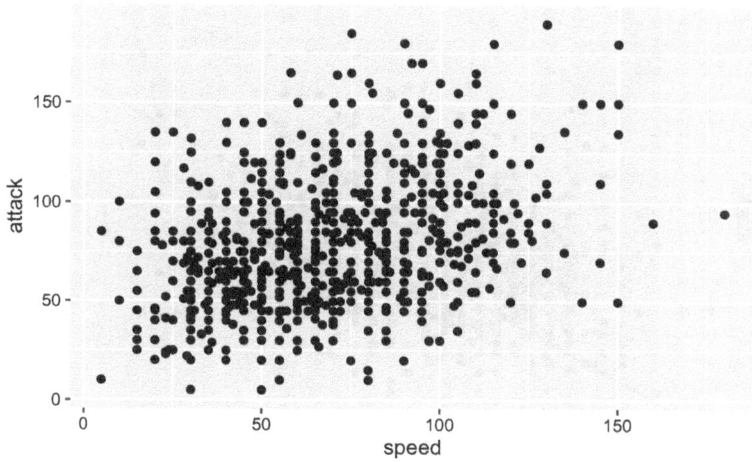

Fig. 6.7 Scatter plot showing the relationship between speed and attack stats

in the study of games, we may want to see if the designers of Pokémon followed this convention. Players who play a lot of similar games may expect their domain knowledge to transfer to games in the same genre. This can help us understand who game manufacturers believe their audiences are and whom they want to cater to.

But, how exactly do we look at attack and speed together? One of the first things that we can do is create a scatter plot showing the relationship between attack and speed.

```
# Examine correlation between speed and attack stats
pokemon |>
  ggplot(aes(x = speed, y = attack)) +
  geom_point()
```

Figure 6.7 suggests our hypothesis that attack and speed go up together seems to hold for most Pokemon. If you recall from our chapter on the grammar of graphics, we can also add a "line of best fit" using the `geom_abline()` function.

```
# Visualize linear relationship between speed and attack with regression
↪ line
pokemon |>
  ggplot(aes(x = speed, y = attack)) +
  geom_point() +
  geom_abline()
```

The line in Fig. 6.8 shows an upward trend where attack increases with speed. Maybe we were wrong about the idea that the Pokémon series does not have conventional roles since even those games that follow role-playing tropes have characters whose attack goes up with speed. However, before we reject our

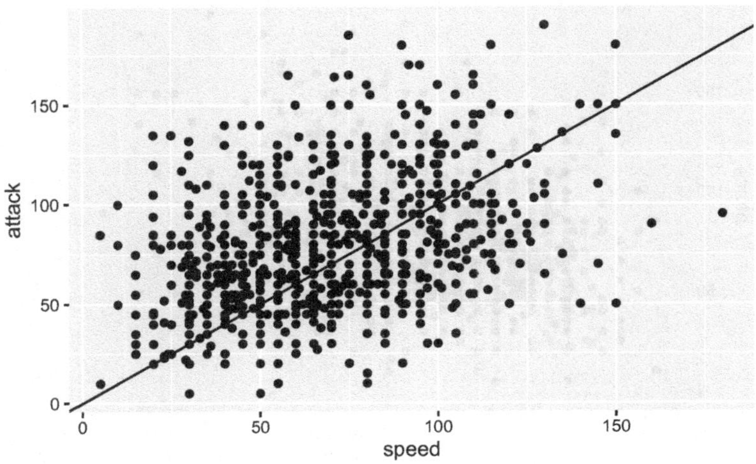

Fig. 6.8 Scatter plot of speed vs. attack with a default reference line

hypothesis that the designers of the games divided their characters into broad categories of strong and weak, we should see if these Pokémon also have low defense. In role-playing games, characters that deal a lot of damage often have low hit points to stop them from being particularly powerful. Let's go ahead and see if attack also goes up with hp.

```
# Examine correlation between HP and attack stats with regression line
pokemon |>
  ggplot(aes(x = hp, y = attack)) +
  geom_point() +
  geom_abline()
```

Figure 6.9 shows that attack also increases with HP. Thus, we can, for now, stick with our thesis that some Pokémon are simply more powerful across the board while others are weak.

We have determined that the relationship between attack and speed along with attack and hp trends upward through the line of best fit. But what exactly does this line represent and how is it computed? What insights can it provide us? To answer these questions, we need to take a deeper dive into linear regression.

6.4.1 Looking at the Line of Best Fit

While we have referred to it as the line of best fit, we can also call it the regression line. But how do we determine this line? To compute it, we begin by taking a predictor variable and plotting it on the x-axis and the outcome variable on

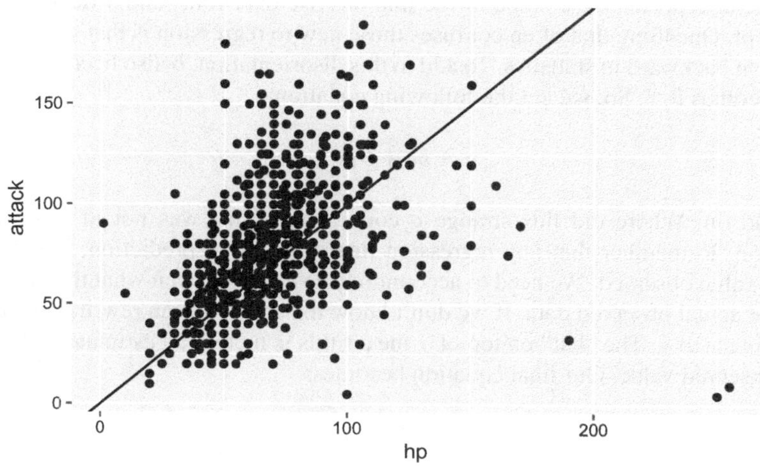

Fig. 6.9 Scatter plot showing the relationship between HP and attack stats with a default reference line

the y-axis. Next, a horizontal straight line is drawn through the average of the outcome variable. Finally, we calculate the distance between each point and the line. These distances represent the "residuals": the value between the line and the actual observed point.

Using an average line is a good first step, and this makes intuitive sense. If we have no other way to predict the outcome variable, we would assume that the majority would be average. However, can we get a measure of just how good of a predictor this is? One way to do this is simply to add up all the "residual" distances, but this presents a problem. Some distances are negative (below the line of average), while others are positive (above the line of average). We could solve this two ways: First, we could get the absolute value of each residual; or second, we could square the values. This second approach is usually the one taken and is more popularly known as "ordinary least squares regression."

Ordinary least squares regression provides us a value that we can use to "optimize" our model. We can rotate the line and recalculate its value to see which line provides us the least squared sum of the residuals. Fortunately, we do not have to do this iterative process ourselves. R calculates it for us, and at the end, a line and its equation are created.

Calculating a line's equation is beneficial, because it provides us a statistical measure for how different variables in our data interact. From grade school, you may remember that the equation for a line is the following:

$$y = mx + b$$

Here m represents the slope of the line or "rise over run," and b denotes the y-intercept. One thing that often confuses those new to regression is that we write this equation backward in statistics. To add to this disorientation, b also becomes b_0, and mx becomes b_1x. So, we get the following equation:

$$y = b_0 + b_1x + \epsilon$$

Hold on. Where did this strange ϵ come from? This was not in our original equation. Remember that any regression line is simply a prediction; it is not the actual value observed. We need to account for the error between what this line gives and the actual observed data. If we don't know the error, we can rewrite this line by changing y to $\hat{y}$. The "hat" on top of y means this is merely an estimate and not the true observed value. Our final equation becomes:

$$\hat{y} = b_0 + b_1x$$

6.4.2 Getting the Equation of a Regression Line

Now that we have a good understanding of how regression works, we can go back to analyzing our dataset. Visualization techniques are useful for communicating to audiences unfamiliar with statistics. However, if you are interested in publishing your research, you will need to provide exact numerical values. So, if we are looking at attack and speed, we want to know exactly, on average, how much we can expect the value of attack to increase with a one-point increase in speed.

Luckily, we can use the `lm()` function to get the equation of our line. This function follows R's formula syntax, which separates an outcome variable from a predictor variable using the tilde (~) operator. The outcome variable, in this case, will be attack, and we will try to predict how much it changes based on speed. Thus, we can write our formula in R as `attack ~ speed`. We will pass the data we are using to the `data` argument of the model. Finally, we will get the results of the model by passing the resulting model to the `summary()` function.

```
# Calculate linear regression model predicting attack from speed
lm_1 <- lm(attack ~ speed, data = pokemon)

# Display regression coefficients and model statistics
summary(lm_1)
```

```
Call:
lm(formula = attack ~ speed, data = pokemon)

Residuals:
    Min      1Q  Median      3Q     Max
```

```
-73.993 -21.548  -3.786  18.067 103.136

Coefficients:
            Estimate Std. Error t value Pr(>|t|)
(Intercept) 49.92848    2.71203   18.41   <2e-16 ***
speed        0.42580    0.03655   11.65   <2e-16 ***
---
Signif. codes:  0 '***' 0.001 '**' 0.01 '*' 0.05 '.' 0.1 ' ' 1

Residual standard error: 30.02 on 798 degrees of freedom
Multiple R-squared:  0.1453,    Adjusted R-squared:  0.1443
F-statistic: 135.7 on 1 and 798 DF,  p-value: < 2.2e-16
```

The output summary provides a plethora of statistical insights, each offering a different perspective on the data. For the moment, let's concentrate on the coefficients, which are the numerical pillars of our regression equation. These coefficients are the result of the regression algorithm finding the "best fit" line through our data points, and they hold the key to understanding the relationship between our variables.

Starting with the intercept, this is a constant term in the regression equation that provides a baseline value for the predicted variable—in this case, the attack stat of a Pokémon. When all other predictor variables are set to zero, the intercept gives us the expected value for the outcome variable. In a practical sense, the concept of a Pokémon with zero speed doesn't align with the reality of the game mechanics, making the intercept more of a mathematical artifact than a real-world indicator. It sets the starting point of the regression line on the y-axis, and although it's necessary for the equation to be complete, it often doesn't carry a meaningful interpretation on its own when considering the predictor variables.

On the other hand, the coefficient for the speed variable has a more tangible interpretation. The positive value of 0.4258 indicates a direct and linear relationship between speed and attack: As a Pokémon's speed increases by one unit, its attack is expected to increase by approximately 0.4258 units, holding all other variables constant. This relationship is at the heart of linear regression, allowing us to predict changes in the dependent variable (attack) based on changes in the independent variable (speed).

$$\text{attack} = \alpha + \beta_1(\text{speed})$$

Entering in the values we received, we get the following equation:

$$\widehat{\text{attack}} = 49.92848 + 0.4258(\text{speed})$$

When we distill the complexities of a dataset into a regression model, we are effectively creating a simplified representation of the relationships within the data. It's crucial to understand that the resulting equation, while powerful, doesn't encapsulate every nuance and detail of the underlying dynamics. In the context of a Pokémon's attack level, numerous factors beyond speed could play a significant role.

Characteristics such as the Pokémon's type—a fundamental aspect of gameplay that influences a Pokémon's strengths and weaknesses—may significantly impact its attack capability.

The reality is that our world, including the virtual world of Pokémon, operates on a web of interconnected factors, and a single-variable model can only reveal so much. Here is where the R^2 value, also known as the coefficient of determination, serves as an essential tool. It provides us with a proportion, ranging from 0 to 1, indicating how much of the total variance in our dependent variable—Pokémon attack in this instance—can be accounted for by the independent variables. An R^2 value of 1 would mean that our model explains all the variance in the attack variable perfectly, a scenario that is theoretically ideal but practically unattainable with real-world data. Conversely, an R^2 value of 0 implies that the model does not explain any of the variance, rendering it ineffective for both descriptive and predictive purposes.

In the statistical analysis of our example, the Multiple R^2 value, often denoted as R^2, is calculated to be approximately 0.1453. This figure translates to about 14.53% when expressed as a percentage. What this tells us is that when considering the entire variability observed in the attack stats across all Pokémon in our dataset, our current model, which only includes speed as a predictor, accounts for a modest 14.53% of that total variance. Put another way, the changes in speed among Pokémon can explain 14.53% of the differences in their attack stats. This leaves a substantial amount, over 85%, of the variance unexplained, which suggests that there are other factors at play influencing a Pokémon's attack that are not captured by speed alone.

The term Multiple R^2 can indeed cause some confusion, as it suggests the use of multiple predictors. In reality, this is a bit of a misnomer when we only have one predictor, as in our current model. However, R uses this terminology consistently, because the `lm()` function is built to handle multiple predictors, and thus the software anticipates that possibility. The Multiple R^2 value will still be accurate in the single predictor case—it just leaves room for the inclusion of more variables.

When additional predictors are included in the model, the Multiple R^2 value will adapt to reflect the proportion of variance explained by the collective predictors. For instance, if we were to add Pokémon type or legendary status as additional predictors, the Multiple R^2 value would increase if these variables contribute significantly to explaining the variance in attack.

An important caveat to consider is the Adjusted R^2 value, which modifies the Multiple R^2 value to account for the number of predictors used. It introduces a penalty for adding more variables to the model, balancing the trade-off between complexity and explanatory power. This adjustment is crucial, because adding more variables to a model will always increase the Multiple R^2 value, but this doesn't necessarily mean we are achieving a better model. The Adjusted R^2 value, therefore, offers a more nuanced and arguably more honest view of a model's explanatory ability. We will use this measure soon as we start adding additional predictors.

6.5 Exploring P-Values

Each coefficient derived from our regression analysis comes with an associated p-value, denoted in the output as `Pr(>|t|)`. This p-value is a fundamental concept in statistical hypothesis testing and is often misunderstood, even within the scientific community (Benjamin and Berger 2019; Greenland 2019; Wellek 2017). Contrary to what many believe, a p-value does not quantify the probability that the null hypothesis is false or the likelihood that the alternative hypothesis is true. Rather, a p-value measures the probability of observing a test statistic as extreme as, or more extreme than, the result obtained by the analysis, assuming that the null hypothesis is correct. In simpler terms, it provides a measure of how surprising the observed data would be if there were truly no effect or association—in our context, if there were no relationship between speed and a Pokémon's attack.

The standard threshold for determining statistical significance is often set at a p-value of less than 0.05. This cutoff, also known as alpha, is somewhat arbitrary but widely adopted in practice. If a p-value falls below this threshold, it is commonly interpreted as evidence strong enough to reject the null hypothesis and suggest that there is a statistically significant relationship between the variables in question. Conversely, a p-value greater than 0.05 implies that the evidence at hand does not suffice to discard the null hypothesis. It's important to note that this does not confirm the null hypothesis. Rather, it suggests the data does not provide strong evidence to conclude there is an effect at the given alpha level.

In the example of our regression model, both coefficients are accompanied by p-values that are below the 0.05 threshold, indicating that the relationships they represent are statistically significant. This statistical significance suggests that the predictors in question have a meaningful association with the outcome variable, justifying further consideration in our analysis and interpretation. However, even with our p-values, it is imperative to check the underlying assumptions of the linear regression model. Ignoring these checks can lead to overconfidence in a model's predictive power and are potentially misleading.

6.6 Checking Assumptions of Linear Models

Regression is powerful, but it is common for individuals to jump directly into posting an equation and its associated p-values without checking that all the assumptions required of a linear regression model are met.

For linear models, there are four key assumptions that should be valid for the model. The first is the assumption of linearity (Osborne and Waters 2019). This is the bedrock of linear regression, and it underpins the expectation that a change in the predictor variable will result in a proportional change in the response variable. This proportionality is what we mean when we describe a relationship as linear. As James et al. caution, if the association between the variables is not linear, our model's predictions will be off, and the insights we glean from the regression line may be misleading. For instance, if the relationship between the attack and speed of Pokémon is actually curved or follows a more complex pattern, a simple linear model may not capture the essence of this relationship.

However, nonlinearity does not automatically preclude the use of linear models. Through the application of various transformations—for instance, taking the log, square, or square root of predictors or the outcome variable—we can sometimes linearize relationships and thus satisfy the linearity assumption. This flexibility allows us to model a broader range of data and still avail ourselves of the interpretative simplicity of linear models. The key, however, is that our model should be linear in terms of its coefficients, even if transformations are applied to the variables themselves.

The second presumption is that residuals are normally distributed. When we assume normality of residuals, we are effectively saying that while individual data points may not perfectly align with the regression line, their deviations—when compiled on a large scale—will distribute symmetrically around the line. Why is this important? For one, it underpins the validity of our confidence intervals and hypothesis tests. Many of the statistical tests we rely on, including the t-tests for assessing the significance of individual regression coefficients, are based on the assumption that the data conforms to a normal distribution. When residuals are normally distributed, it gives us confidence that the p-values we use to determine statistical significance are reliable. When residuals do not follow a normal distribution, it can indicate problems with the model, such as omitted variables that are important predictors of the response variable, or it could suggest that we have a nonlinear relationship that hasn't been properly accounted for in the model. Non-normality can also indicate outliers in the data that exert an undue influence on the model's parameters.

The third presumption for linear regression is homoscedasticity. Homoscedasticity ensures that the spread or dispersion of the residuals is consistent across all levels of the predictor variables. When this assumption is violated, we see patterns of increasing or decreasing spread in the residuals as we move along the range of predictor variables (heteroscedasticity). This signals that our model might not be weighing all data points equally.

The final foundational assumption of linear models is the independence of residuals. Each data point should contribute fresh information to the model. If the data points are related—for example, if we're looking at repeated measures from the same subject or clustered observations from related individuals—these relationships could influence the residuals in a way that does not reflect the assumption of independence. This interconnectedness can introduce bias, since the model will treat interrelated data points as though they are contributing unique information, leading to overconfidence in the model's predictions. It is especially pertinent to consider this assumption in datasets with a temporal or spatial component, such as time-series data or data clustered by physical proximity or social connection. Such data structures can introduce autocorrelation, where the residuals are correlated with each other and violate the assumption of independence.

Various statistical tests have been created to check for these assumptions, but the easiest technique is to look at model results visually. One of the best packages for doing this is the "performance" package. Created by the Core R Team, it is part of the "easystats" library of packages (Lüdecke et al. 2022). To use it, we simply pass our model to its `check_model()` function.

````
```{r}
# Uncomment to install 'performance' package
# install.packages("performance")
# Load package for regression diagnostics
library(performance)

# Validate regression model assumptions with formatted output
check_model(lm_1)
```
````

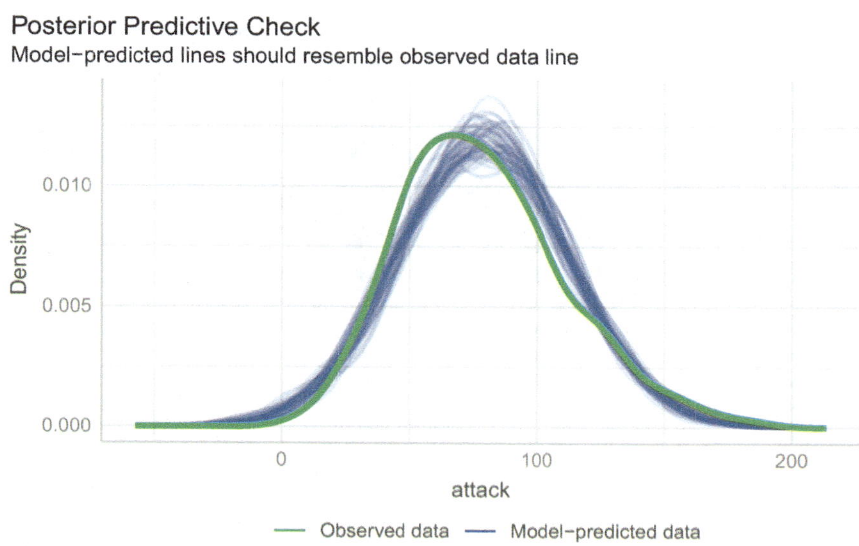

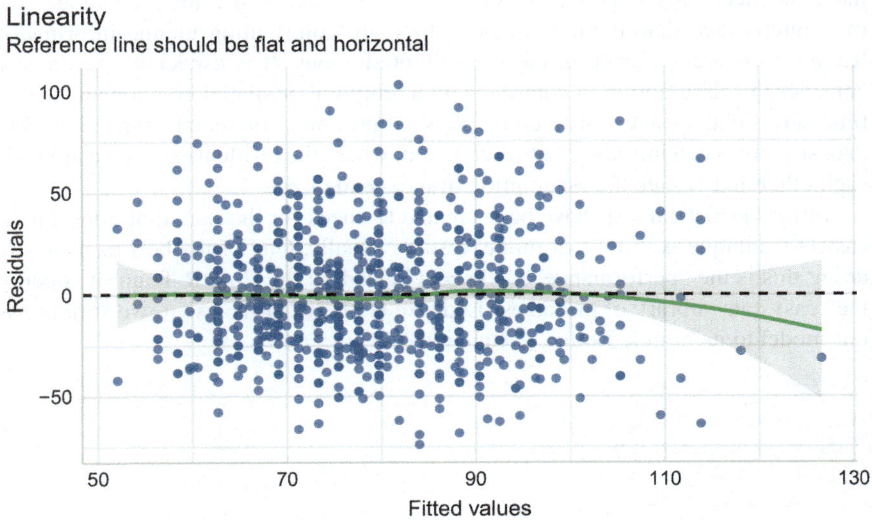
Linearity
Reference line should be flat and horizontal
100
50
0
−50
Residuals
50
70
90
110
130
Fitted values

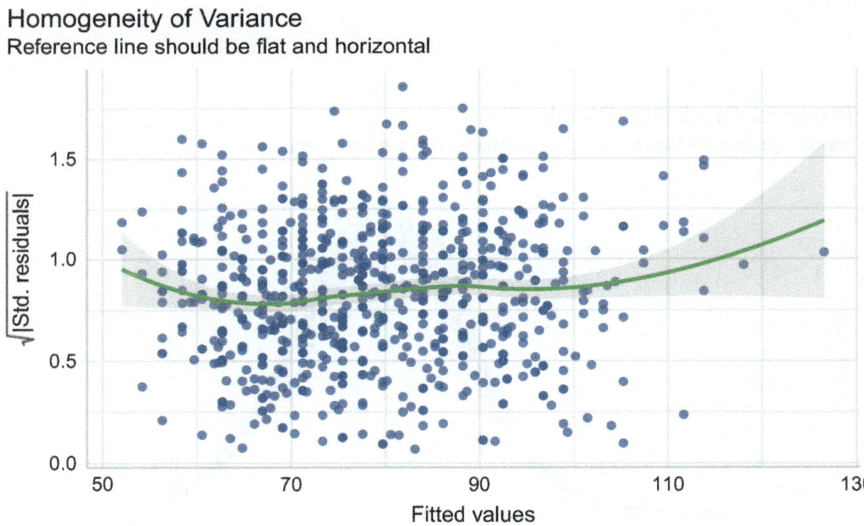
Homogeneity of Variance
Reference line should be flat and horizontal
1.5
1.0
0.5
0.0
√|Std. residuals|
50
70
90
110
130
Fitted values

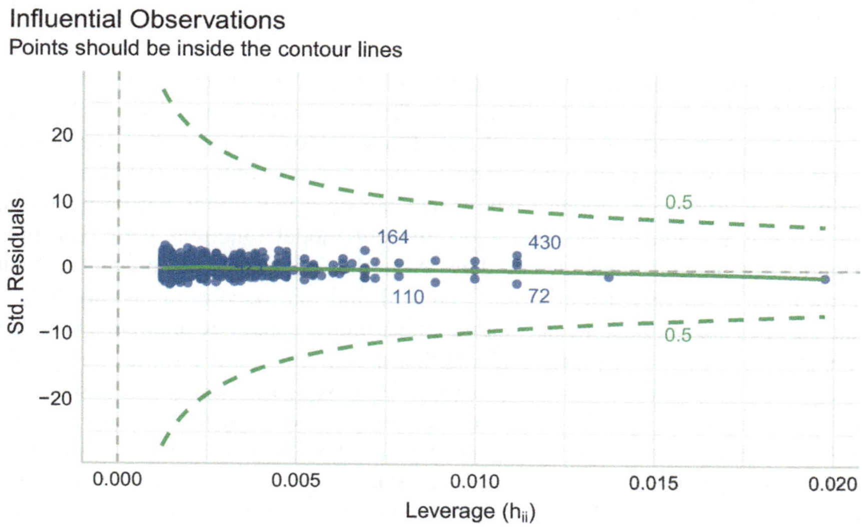

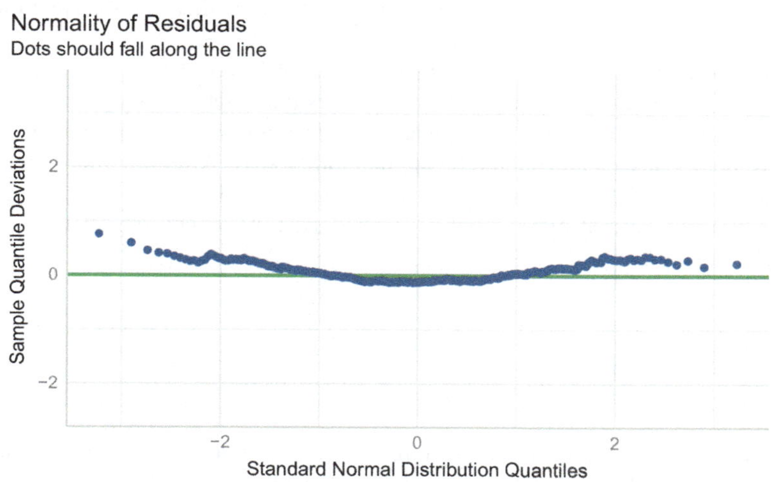

The visual diagnostics can help us interpret whether the assumptions of the linear regression model are being met. In this case, we have a strong model, which means we can be fairly certain that the coefficient estimates are reliable. However, do not worry if your dataset does not meet all assumptions perfectly. You just want to make sure there is nothing completely outrageous unless you have very small sample sizes.

6.6.1 Checking Skew

Let's see if there is a relationship between the defense and attack of each Pokémon. Just as we did before, we will use the `lm()` function to get the results and check the assumptions of our linear model with the `check_model()` function. Doing this can be tedious, however, as we often don't know if our assumptions will be met until we finish building our model. Luckily, we can do a few things before beginning our analysis to increase the likelihood that our model functions well. The most important of these is checking for skew. Note that skew does not prevent us from performing linear regression. Often, you will find that people mistakenly believe that skewness of our data means that the linear model cannot be applied. This is not true. It is the skewness of the residuals that is important, not the skewness of the predictor or outcome variables themselves. However, skewness in the distribution of either the predictor or the response variable is more likely to lead to a violation of the normality of residuals or the linearity assumption. We can identify skewness by visualizing the distribution of each variable—typically through the creation of histograms. If we do find that our data is skewed, we might consider applying a transformation to the data, such as taking the logarithm, the square root, or some other nonlinear transformation, which can help to normalize the distribution and stabilize the variance across the data.

Transforming skewed data brings us closer to satisfying the assumptions necessary for linear regression, which in turn enhances the reliability and interpretability of our results. It's worth noting that while transformations can help in many cases, they are not a panacea and should be applied judiciously. After any transformation, it's essential to recheck the distribution of the transformed data to ensure that it now approximates a normal distribution more closely. Let's check the skew of attack through a histogram.

```
# Analyze distribution of attack values for normality
pokemon |>
  ggplot(aes(x = attack)) +
  geom_histogram()
```

Figure 6.10 shows a slight asymmetry leaning toward the right, which is indicative of a right skew. This skewness is shaped, in part, by the presence of a few unusually strong outliers—Pokémon with exceptionally high attack stats that deviate significantly from the norm. Despite their presence, these outliers are not extreme enough to raise concerns of significantly distorting our analytical outcomes or predictive models. Their influence appears to be contained such that the integrity of our subsequent analyses remains intact.

We will now turn our attention to another critical stat—defense. As with attack, understanding the distribution of defense values is vital for constructing reliable models. Defense is a crucial factor in battle outcomes, and it stands to reason that

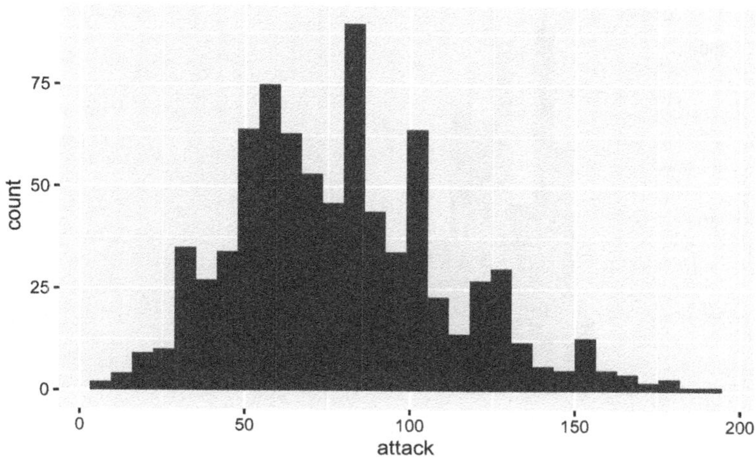

Fig. 6.10 Histogram showing the distribution of attack values

its distribution could have a different shape and spread compared to attack values. Recognizing peculiarities in the defense statistics, such as skewness or the presence of outliers, is essential for the same reasons we scrutinize attack—ensuring that our regression models are built on solid, representative grounds.

```
# Analyze distribution of defense values for normality
pokemon |>
  ggplot(aes(x = defense)) +
  geom_histogram()
```

Figure 6.11 reveals a pattern similar to what we noticed with attack—a slight skew, this time also to the right. This suggests that while there are indeed Pokémon with defense values that are higher than the general population, these cases are not so numerous or extreme as to skew the entire distribution dramatically.

We can now go ahead and create a linear regression model looking at defense and attack. Let's begin with a simple visualization to see if the trend is positive or negative.

```
# Examine correlation between defense and attack with regression line
pokemon |>
  ggplot(aes(x = defense, y = attack)) +
  geom_point() +
  geom_abline()
```

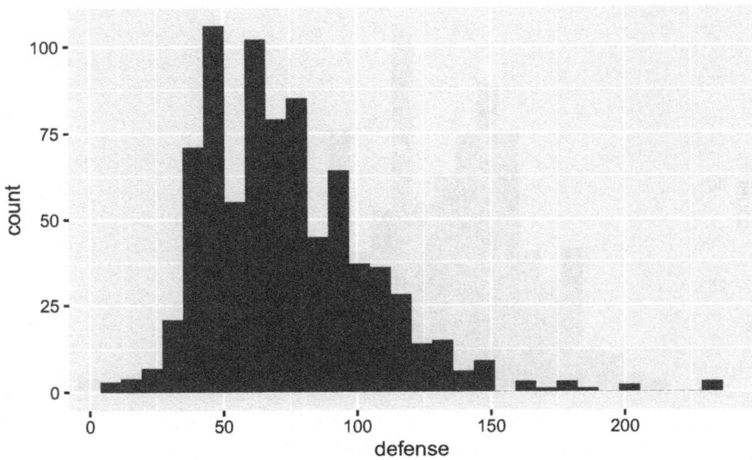

Fig. 6.11 Histogram showing the distribution of defense values

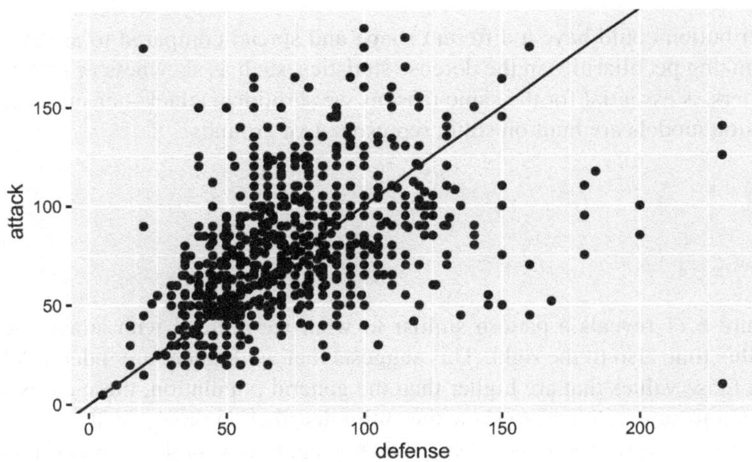

Fig. 6.12 Scatter plot showing the relationship between defense and attack stats with a default reference line

The line in Fig. 6.12 trends upward. But just how much does defense increase with attack exactly? Remember that to get this value, we can get the equation for the line using the `lm()` and `summary()` functions.

```
# Calculate linear regression model predicting attack from defense
lm_2 <- lm(attack ~ defense, data = pokemon)

# Display regression coefficients and model statistics
summary(lm_2)
```

```
Call:
lm(formula = attack ~ defense, data = pokemon)

Residuals:
     Min       1Q   Median       3Q      Max
-140.304  -19.203   -4.039   16.234  125.584

Coefficients:
            Estimate Std. Error t value Pr(>|t|)
(Intercept) 45.28420    2.65383   17.06   <2e-16 ***
defense      0.45661    0.03311   13.79   <2e-16 ***
---
Signif. codes:  0 '***' 0.001 '**' 0.01 '*' 0.05 '.' 0.1 ' ' 1

Residual standard error: 29.19 on 798 degrees of freedom
Multiple R-squared:  0.1924,	Adjusted R-squared:  0.1914
F-statistic: 190.2 on 1 and 798 DF,  p-value: < 2.2e-16
```

Looking at the Multiple R^2 value, we see that roughly 19.24% of attack can be explained by defense. We also see that we have a positive value for defense that is statistically significant. When defense goes up 1 point, attack goes up 0.45661. Thus, the equation for our regression line is:

$$\widehat{\text{attack}} = 45.28420 + 0.45661(\text{defense})$$

Let's go ahead and check the assumptions.

```
```{r}
# Validate regression model assumptions with formatted output
check_model(lm_2)
```
```

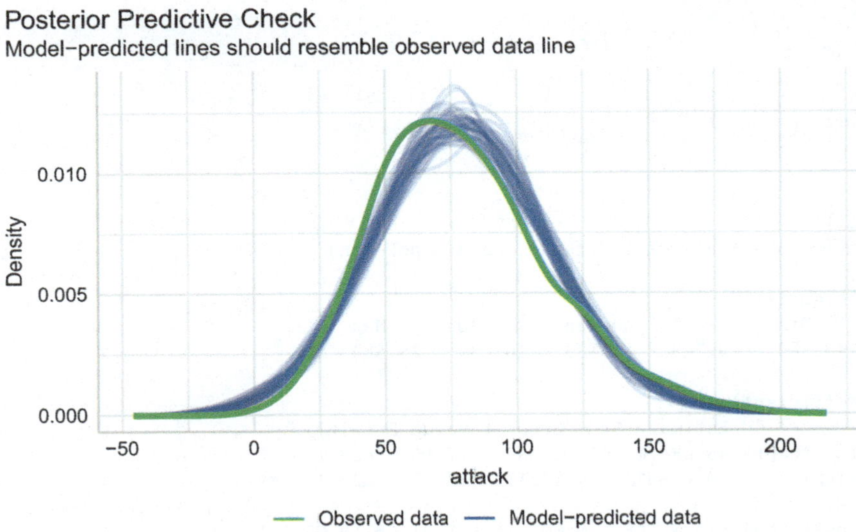
Posterior Predictive Check
Model−predicted lines should resemble observed data line
Density
0.010
0.005
0.000
−50
0
50
100
150
200
attack
Observed data
Model−predicted data

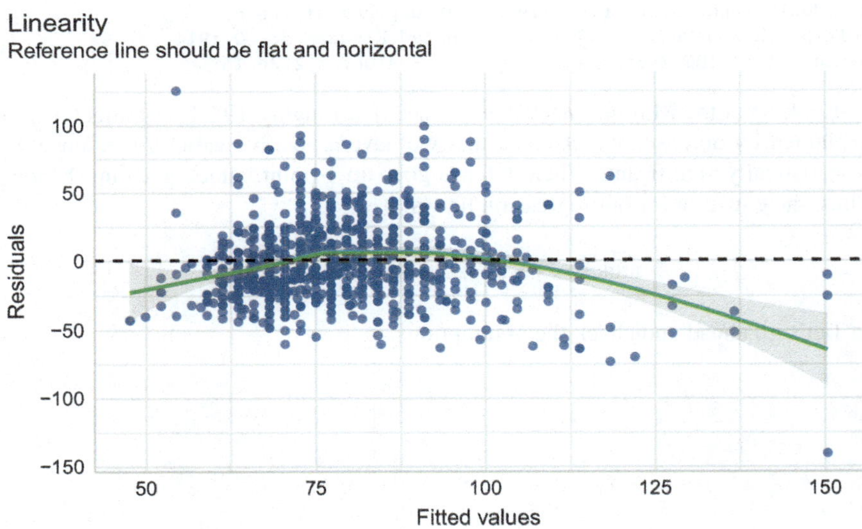
Linearity
Reference line should be flat and horizontal
Residuals
100
50
0
−50
−100
−150
50
75
100
125
150
Fitted values

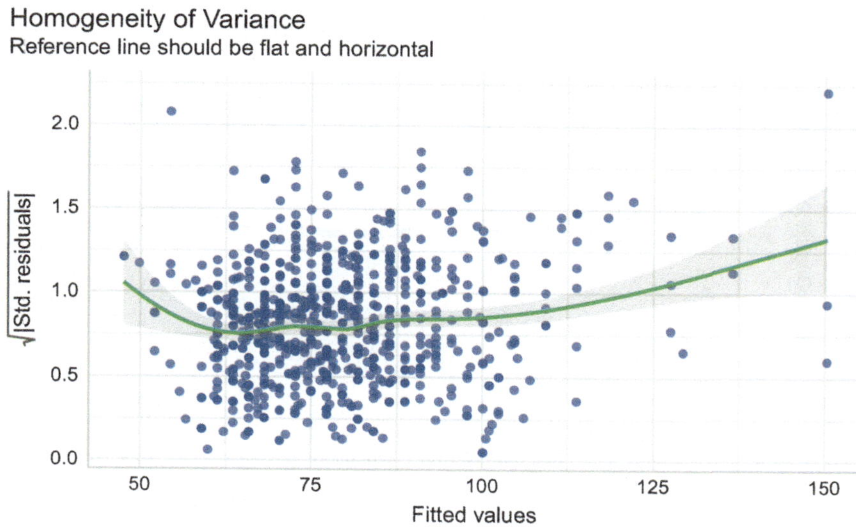
Homogeneity of Variance
Reference line should be flat and horizontal
√|Std. residuals|
2.0
1.5
1.0
0.5
0.0
50
75
100
125
150
Fitted values

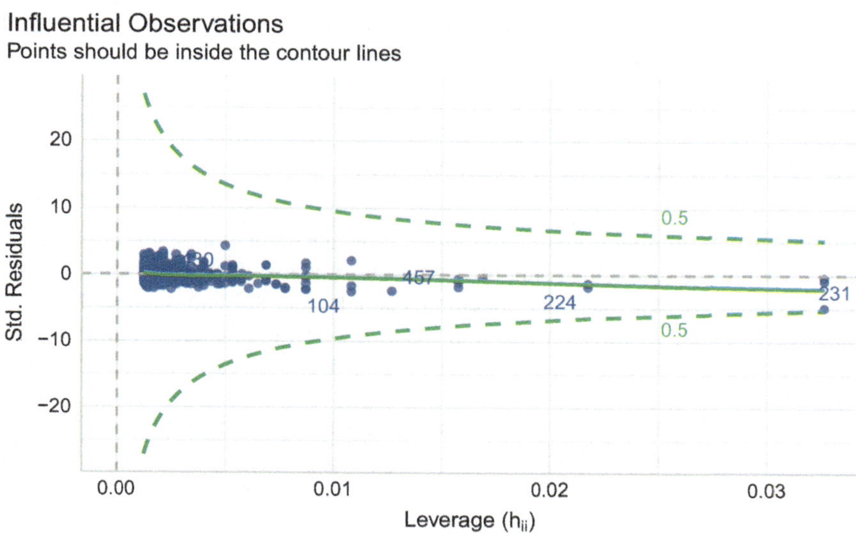
Influential Observations
Points should be inside the contour lines
Std. Residuals
20
10
0
−10
−20
0.5
0.5
104
457
224
231
0.00
0.01
0.02
0.03
Leverage (hii)

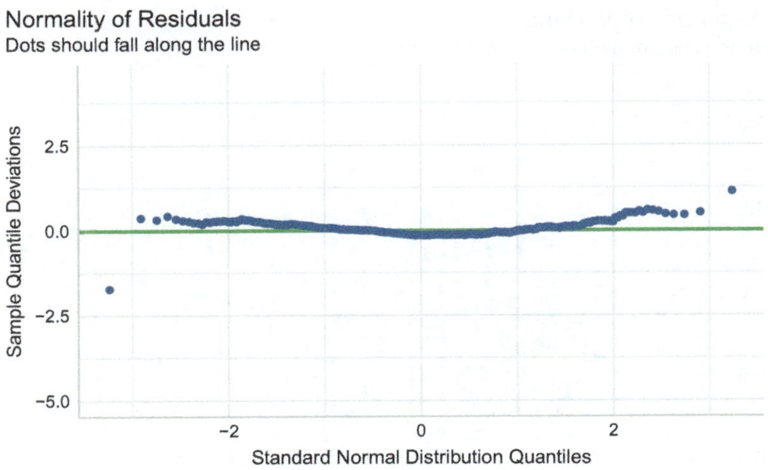

Based on the diagnostic analysis, it is safe to say that our formula can be useful for prediction. These results provide more evidence that certain Pokémon are simply stronger than others across the board rather than only for those characteristics that may contribute to a particular role (damage dealer, defensive, etc.).

6.7 Multivariate Linear Regression

Rather than determining if Pokémon follows typical role-playing game conventions, maybe we are interested in seeing how different statistics interact with one another more broadly. We saw that each Pokémon's attack went up with the speed, hp, and defense. However, we have only been measuring one variable against another. This can be laborious after a while, and it is also likely that more than one statistic relates to attack.

Luckily, adding additional predictors to a linear regression model in R is easy. We simply place these additional variables to the right of the ~ and separate them with a plus sign in our call to `lm()`. When more than one predictor variable is used, we refer to this as multivariate linear regression.

Just as before, we will begin by checking the skew using histograms. It is fairly tedious having to create a separate histogram for each of the values. Fortunately, we can get a quick glance at the distribution and summary statistics of each column in our data using the "skimr" package and its accompanying `skim()` function. Note that you may need to adjust your window size to see the accompanying histograms (Table 6.1).

Table 6.1 Data summary

Name	pokemon
Number of rows	800
Number of columns	12
Column type frequency:	
character	3
factor	1
logical	1
numeric	7
Group variables	None

```
# Install skimr package if necessary
# install.packages("skimr")

# Load package for comprehensive data summaries
library(skimr)

# Generate statistical summaries and visualizations of all variables
skim(pokemon)
```

Variable type: character

skim_variable	n_missing	complete_rate	min	max	empty	n_unique	whitespace
name	1	1.00	3	25	0	799	0
type_1	0	1.00	3	8	0	18	0
type_2	386	0.52	3	8	0	18	0

Variable type: factor

skim_variable	n_missing	complete_rate	ordered	n_unique	top_counts
generation	0	1	FALSE	6	1: 166, 5: 165, 3: 160, 4: 121

Variable type: logical

skim_variable	n_missing	complete_rate	mean	count
legendary	0	1	0.08	FAL: 735, TRU: 65

Variable type: numeric

skim_variable	n_missing	complete_rate	mean	sd	p0	p25	p50	p75	p100	hist
id	0	1	400.50	231.08	1	200.75	400.5	600.25	800	
hp	0	1	69.26	25.53	1	50.00	65.0	80.00	255	
attack	0	1	79.00	32.46	5	55.00	75.0	100.00	190	
defense	0	1	73.84	31.18	5	50.00	70.0	90.00	230	
sp_attack	0	1	72.82	32.72	10	49.75	65.0	95.00	194	
sp_def	0	1	71.90	27.83	20	50.00	70.0	90.00	230	
speed	0	1	68.28	29.06	5	45.00	65.0	90.00	180	

We see that, like attack and speed data, most variables have a slight right skew. However, nothing is so extreme that it will impact our model's validity.

We can now create the model itself. We will begin by seeing how speed, hp, and defense relate to attack.

```
# Calculate multiple regression predicting attack from speed, HP, and
↪ defense
lm_3 <- lm(attack ~ speed + hp + defense, data = pokemon)

# Display regression coefficients and model statistics
summary(lm_3)
```

```
Call:
lm(formula = attack ~ speed + hp + defense, data = pokemon)

Residuals:
     Min       1Q   Median       3Q      Max
-105.100  -16.466   -1.985   14.305   98.523

Coefficients:
            Estimate Std. Error t value Pr(>|t|)
(Intercept)  1.46266    3.47762   0.421    0.674
speed        0.36515    0.03109  11.745   <2e-16 ***
hp           0.35188    0.03644   9.656   <2e-16 ***
defense      0.38238    0.02938  13.016   <2e-16 ***
---
Signif. codes:  0 '***' 0.001 '**' 0.01 '*' 0.05 '.' 0.1 ' ' 1

Residual standard error: 25.13 on 796 degrees of freedom
Multiple R-squared:  0.4027,	Adjusted R-squared:  0.4005
F-statistic: 178.9 on 3 and 796 DF,  p-value: < 2.2e-16
```

Most of this output should be familiar. However, there are two important things to note. The first and most obvious is that we have more coefficients. Our regression model equation can thus be written as follows:

$$\widehat{\text{attack}} = 1.46266 + 0.36515(\text{speed}) + 0.35188(\text{hp}) + 0.38238(\text{defense})$$

Second, rather than using the Multiple R^2 value, we need to use the Adjusted R^2 value. As mentioned earlier, when using more than one predictor, Multiple R^2 will always report a higher value. Adjusted R^2 takes this into account and adds a "penalty" for using too many predictors. We can understand it in the same way as the Multiple R^2 value. For our model, 40.05% of the variance in attack can be explained by the variance in speed, hp, and defense.

Before we can report these values, we will use the `check_model()` function again to make sure our assumptions are being met. Remember that in practice, you will rarely have a perfect model.

```
```{r}
# Validate multiple regression model assumptions with formatted output
check_model(lm_3)
```
```

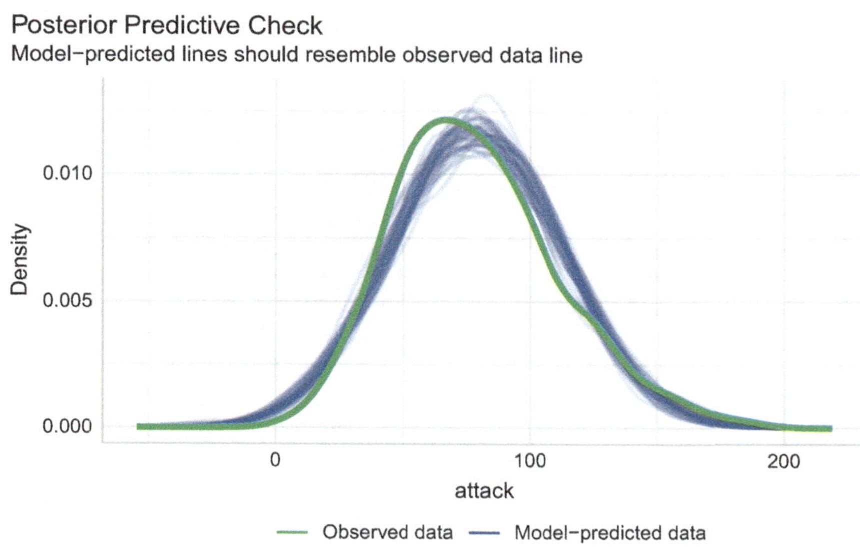

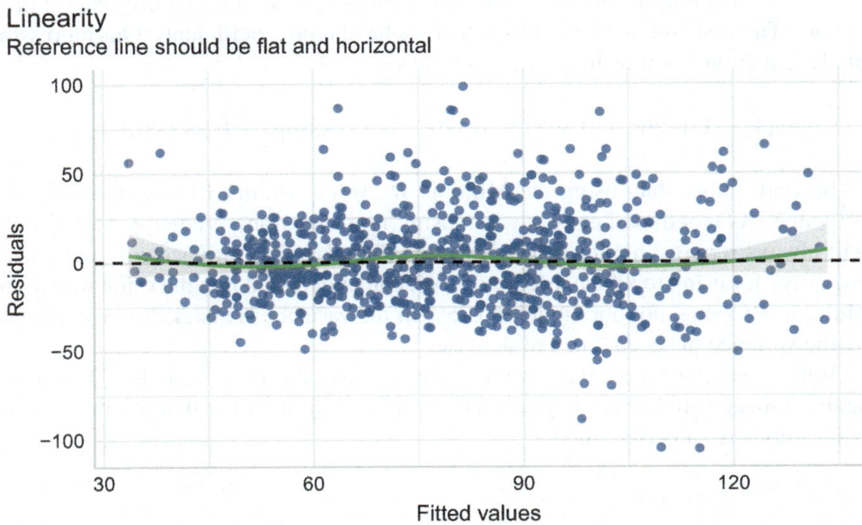
Linearity
Reference line should be flat and horizontal
100
50
0
−50
−100
Residuals
30
60
90
120
Fitted values

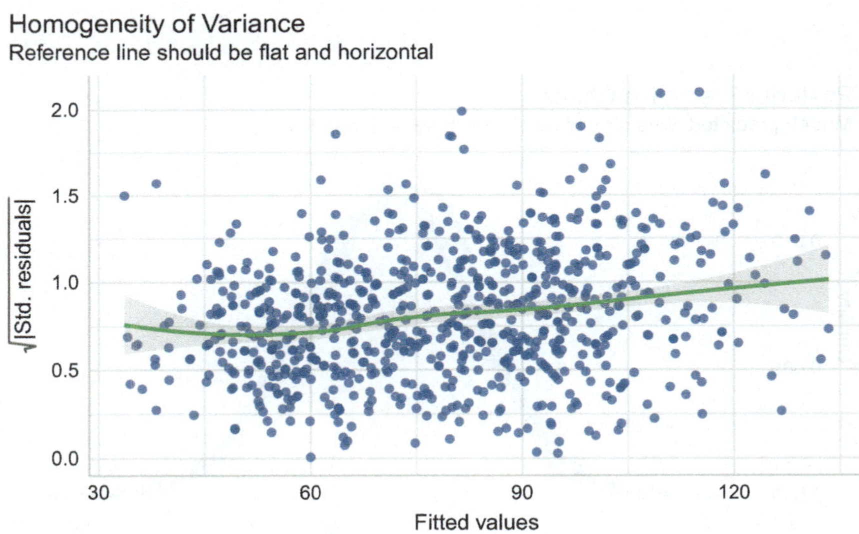
Homogeneity of Variance
Reference line should be flat and horizontal
2.0
1.5
1.0
0.5
0.0
√|Std. residuals|
30
60
90
120
Fitted values

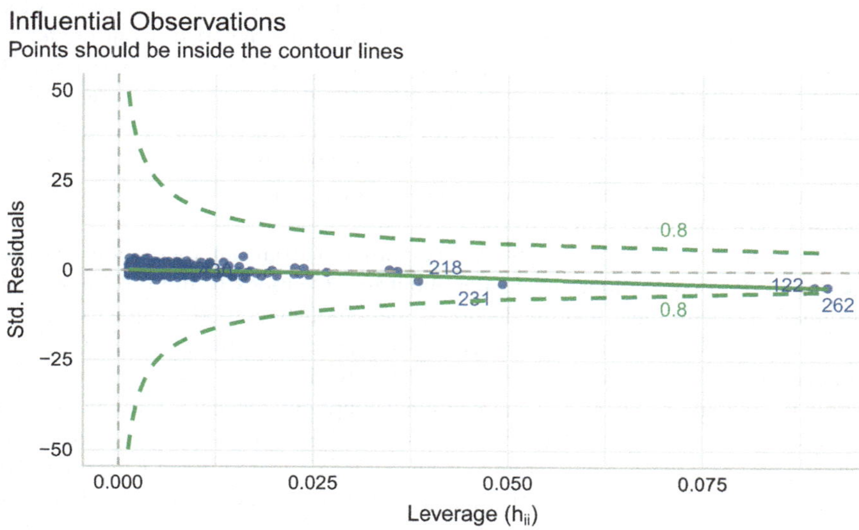

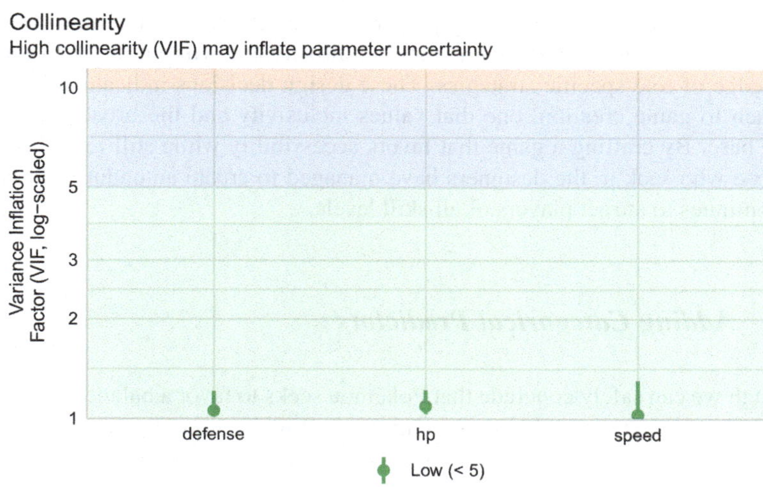

The diagnostic results show that most of the necessary assumptions are satisfactorily met—linearity, homoscedasticity, and the independence of residuals are all within acceptable bounds. However, one critical assumption, the normality of residuals, appears to be violated. This lack of normality in the residuals could be attributed to several factors inherent to the nature of the dataset and the subject matter. Pokémon statistics are designed to capture a diverse array of attributes and abilities that may not necessarily follow a normal distribution in the population. As we have seen, some Pokémon have been designed with exceptional abilities to create a sense of rarity and uniqueness in the game, resulting in outliers that disrupt the normal distribution of residuals. Furthermore, there could be hidden variables that are influential in determining battle outcomes but are not included in our model.

In practice, a slight deviation from normality may not always substantively impair the model's performance, especially with large sample sizes where the central limit theorem tends to minimize the impact of such violations. Nevertheless, it's a point of consideration that analysts must be mindful of, and corrective measures should be engaged if significant non-normality is detected to ensure the robustness of our regression analysis.

Given the data and the insights we've garnered from our regression analysis, we are led to a compelling conclusion about the intentions of the Pokémon game designers. It seems clear that the designers meticulously crafted a game experience that is accessible and inviting to newcomers in the role-playing game (RPG) genre. This design philosophy is reflected in the statistical patterns and trends observed within the Pokémon attributes—an intentional simplicity resonates through the data.

For instance, the observed increase in both attack and defense metrics across various Pokémon suggests a deliberate move away from the traditional RPG archetype of specialized roles—be they tanks, healers, or damage dealers. Instead, Pokémon tend to exhibit balanced attributes, reducing the complexity of understanding and strategizing around diverse character classes. This design choice likely serves to flatten the learning curve, allowing players who might be less familiar with RPG mechanics to engage with the game more intuitively and without the need for deep knowledge of role-specific strategies. These design decisions indicate a thoughtful approach to game creation, one that values inclusivity and the broadening of the player base. By crafting a game that favors accessibility while still retaining depth for those who seek it, the designers have managed to create an enduring franchise that continues to attract players of all skill levels.

6.7.1 Adding Categorical Predictors

Although we can safely conclude that Pokémon seeks to favor a balance of all types, it is worth exploring how the incorporation of all relevant predictors can influence the outcome of our model, which will be particularly important for prediction. However, a problem emerges since much of our data, like type, is categorical. How exactly can we calculate the coefficient value for a categorical variable? The answer is through dummy coding. With dummy coding, we create a binary variable for each category in a given column. In this case, we would create a new variable for each Pokémon type. Underneath, if the Pokémon is that type, we place a 1; otherwise, we place a 0. Fortunately, the `lm()` function will take care of all this for us.

Let's create a model that takes into account how the primary Pokémon type intersects with attack.

```
# Calculate multiple regression including Pokemon type as predictor
lm_4 <- lm(attack ~ hp + defense + speed + type_1, data = pokemon)

# Display regression coefficients and model statistics
summary(lm_4)
```

```
Call:
lm(formula = attack ~ hp + defense + speed + type_1, data = pokemon)

Residuals:
     Min       1Q   Median       3Q      Max
-102.937  -14.112   -1.452   11.400  105.644

Coefficients:
                Estimate Std. Error t value Pr(>|t|)
(Intercept)      3.82221    4.14085   0.923 0.356268
hp               0.35138    0.03630   9.681  < 2e-16 ***
defense          0.33454    0.03166  10.566  < 2e-16 ***
speed            0.38100    0.03118  12.219  < 2e-16 ***
type_1Dark       8.57943    5.21171   1.646 0.100129
type_1Dragon    18.49739    5.23577   3.533 0.000435 ***
type_1Electric -10.11541    4.68243  -2.160 0.031055 *
type_1Fairy     -8.82984    6.54668  -1.349 0.177809
type_1Fighting  21.18172    5.46876   3.873 0.000116 ***
type_1Fire       5.34994    4.44366   1.204 0.228976
type_1Flying   -11.14841   12.40127  -0.899 0.368945
type_1Ghost     -4.35859    5.14063  -0.848 0.396770
type_1Grass     -1.52617    4.08459  -0.374 0.708774
type_1Ground    13.27044    5.16762   2.568 0.010414 *
type_1Ice       -4.44114    5.70711  -0.778 0.436701
type_1Normal    -4.78831    3.87822  -1.235 0.217326
type_1Poison    -0.01838    5.38667  -0.003 0.997279
type_1Psychic  -10.87607    4.35805  -2.496 0.012779 *
type_1Rock      11.05240    4.72142   2.341 0.019488 *
type_1Steel      2.63399    5.70944   0.461 0.644684
type_1Water     -4.52785    3.70736  -1.221 0.222337
---
Signif. codes:  0 '***' 0.001 '**' 0.01 '*' 0.05 '.' 0.1 ' ' 1

Residual standard error: 23.98 on 779 degrees of freedom
Multiple R-squared:  0.468, Adjusted R-squared:  0.4543
F-statistic: 34.26 on 20 and 779 DF,  p-value: < 2.2e-16
```

Our Adjusted R^2 value is 0.4543, which means that the variance in type, hp, speed, defense, and type account for approximately 45.43% of the variance in attack. You will also see that for each of the types, we have a coefficient showing the amount that attack will go up based on if the Pokémon is of that type. Each coefficient for type will be multiplied by a 1 or 0 depending on if the Pokémon is that type. For instance, if a Pokémon was type Dragon, we would multiply 1 by the coefficient of 18.49739. The coefficients for the other types would be multiplied by 0.

You will notice that one type is missing from our analysis. This is used as the intercept and is what the other values are judged from. In the above example, we have an intercept value of 3.82221. This represents the first category alphabetically for the chosen column. In this case, it represents "Bug," so a Pokémon with this type would have an attack of 3.82221 regardless of the other predictors. Taking our Dragon example, we can say that a Pokémon of type Dragon has 18.49739 more attack points on average than the Bug type. Let's check the assumptions again.

```
```{r}
# Validate regression model with categorical predictor (formatted)
check_model(lm_4)
```
```

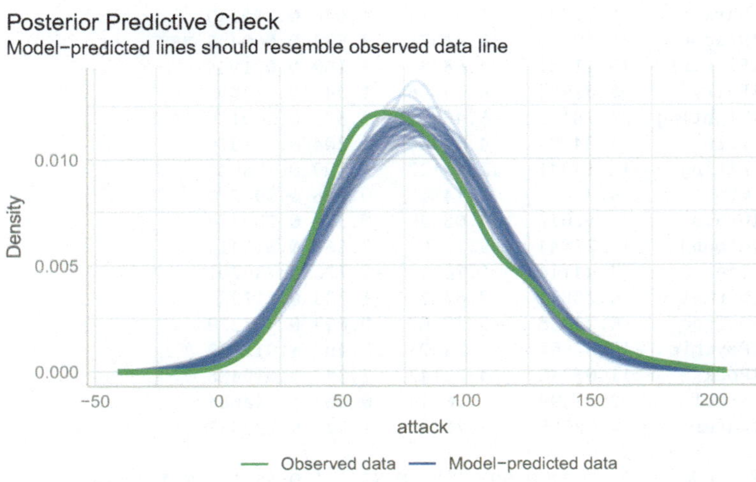

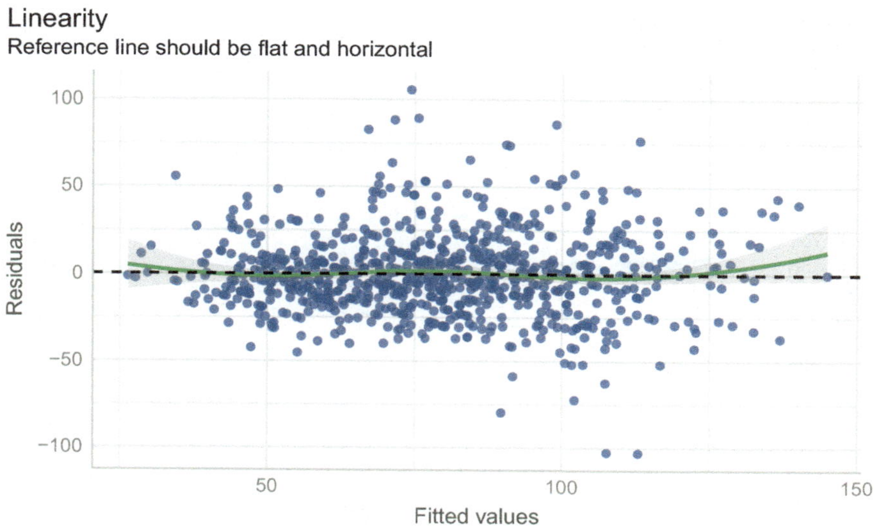
Linearity
Reference line should be flat and horizontal
100
50
0
−50
−100
Residuals
50
100
150
Fitted values

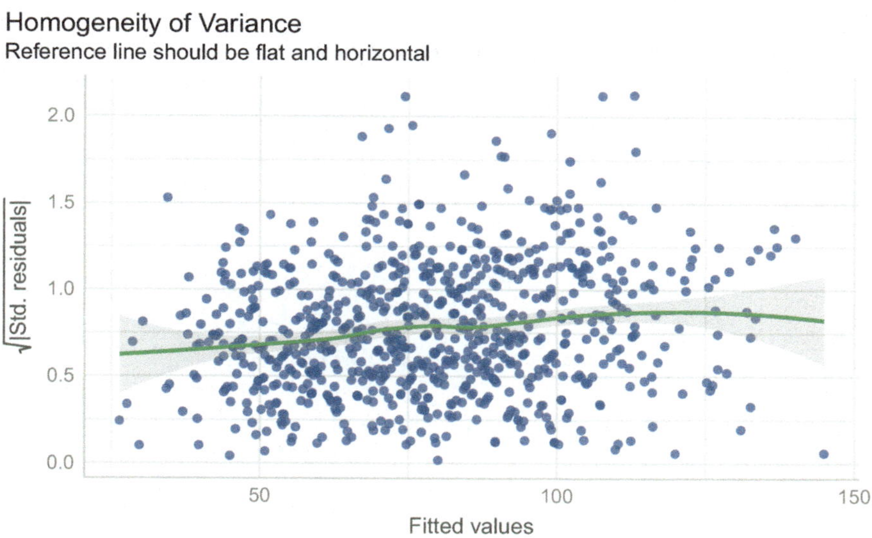
Homogeneity of Variance
Reference line should be flat and horizontal
2.0
1.5
1.0
0.5
0.0
√|Std. residuals|
50
100
150
Fitted values

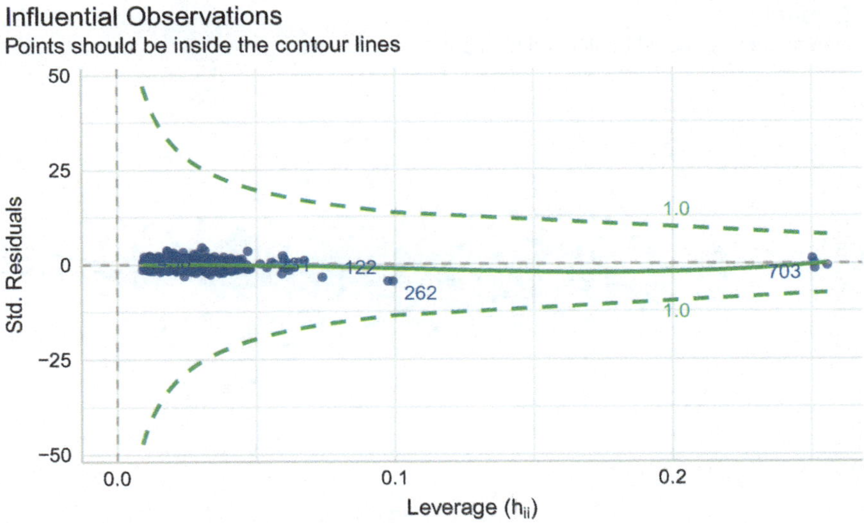

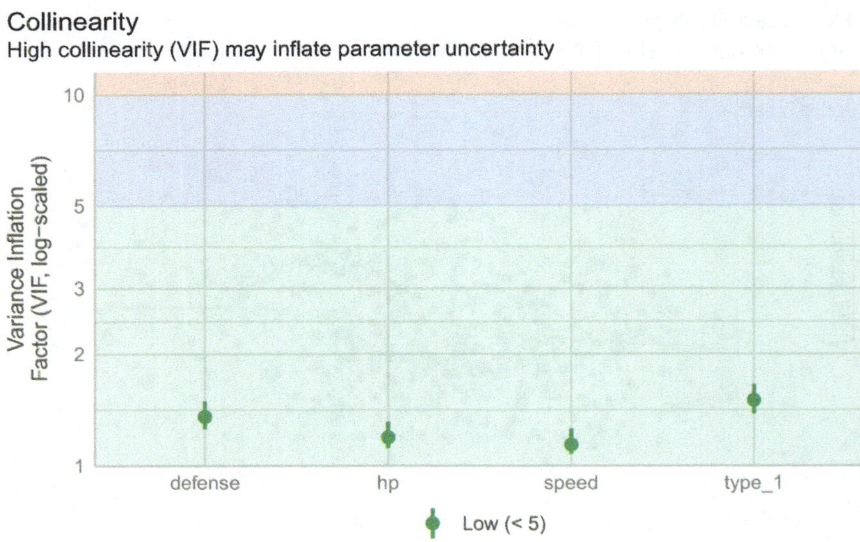

6.7.2 *Adding All Predictors*

Normally, when creating models, you want to use as much data as possible. Rather than typing in each column individually, we can add all predictors to our model using a `.` on the right-hand side of the formula syntax. Remember that we already set the "generation" column to factor; otherwise, the `lm()` function will assume it is numeric. We need to remove the "name" column, since otherwise `lm()` will code a new category for each name when we pass all the predictors. Finally, we will remove the "id" column, since it doesn't add any information we may need for modeling.

Let's create our model and print out its results.

```
# Filter dataset to remove non-predictive columns
pokemon_selected_variables <- pokemon |>
  select(-name, -id)

# Calculate comprehensive regression model using all remaining predictors
lm_5 <- lm(attack ~ ., data = pokemon_selected_variables)

# Display complete regression results
summary(lm_5)
```

```
Call:
lm(formula = attack ~ ., data = pokemon_selected_variables)

Residuals:
    Min      1Q  Median      3Q     Max
-65.031 -12.402  -1.813  11.640  70.879

Coefficients:
                Estimate Std. Error t value Pr(>|t|)
(Intercept)     35.47742   14.76708   2.402 0.016780 *
type_1Dark       1.90263    5.93754   0.320 0.748817
type_1Dragon     7.40145    6.67137   1.109 0.267967
type_1Electric -22.79130    6.76630  -3.368 0.000836 ***
type_1Fairy    -34.12726   16.01251  -2.131 0.033728 *
type_1Fighting   1.73414    9.24769   0.188 0.851355
type_1Fire     -12.89441    6.16689  -2.091 0.037222 *
type_1Flying   -42.49693   17.06010  -2.491 0.013178 *
type_1Ghost    -22.10091    6.23022  -3.547 0.000439 ***
type_1Grass     -9.33323    4.98375  -1.873 0.061898 .
type_1Ground     6.71490    6.28371   1.069 0.285941
type_1Ice      -14.64681    7.82898  -1.871 0.062160 .
type_1Normal    -5.29244    5.14259  -1.029 0.304090
type_1Poison   -18.89417    7.28539  -2.593 0.009882 **
type_1Psychic   -6.11757    6.46825  -0.946 0.344879
type_1Rock      -1.11641    5.32836  -0.210 0.834157
type_1Steel      1.69377    6.57105   0.258 0.796734
type_1Water    -16.14835    4.64613  -3.476 0.000570 ***
type_2Dark       5.04475   13.84977   0.364 0.715882
type_2Dragon   -19.12539   14.22961  -1.344 0.179758
type_2Electric -37.91606   16.15381  -2.347 0.019445 *
type_2Fairy    -36.01454   13.95419  -2.581 0.010240 *
type_2Fighting   2.80806   13.81095   0.203 0.838996
type_2Fire     -29.68984   14.94766  -1.986 0.047747 *
type_2Flying   -26.23918   13.38154  -1.961 0.050651 .
type_2Ghost    -12.31595   14.65176  -0.841 0.401130
```

```
type_2Grass     -14.04591   14.01485  -1.002 0.316898
type_2Ground    -13.50041   13.50354  -1.000 0.318079
type_2Ice       -20.31720   14.49812  -1.401 0.161947
type_2Normal    -43.12978   17.91664  -2.407 0.016565 *
type_2Poison    -23.97526   14.17034  -1.692 0.091505 .
type_2Psychic   -37.80265   13.76846  -2.746 0.006337 **
type_2Rock      -19.64391   14.62544  -1.343 0.180057
type_2Steel      -9.35551   14.03888  -0.666 0.505573
type_2Water     -29.78592   14.14512  -2.106 0.035904 *
hp                0.44435    0.06289   7.066 8.06e-12 ***
defense           0.28973    0.05158   5.617 3.83e-08 ***
sp_attack         0.20239    0.05105   3.964 8.85e-05 ***
sp_def           -0.18258    0.05678  -3.216 0.001416 **
speed             0.31152    0.04906   6.349 6.37e-10 ***
generation2      -5.31992    4.20427  -1.265 0.206543
generation3      -0.58510    3.96238  -0.148 0.882689
generation4      -6.95447    4.03486  -1.724 0.085621 .
generation5      -8.20419    3.87855  -2.115 0.035078 *
generation6       0.57355    4.75514   0.121 0.904061
legendaryTRUE     6.73186    4.58149   1.469 0.142589
---
Signif. codes:  0 '***' 0.001 '**' 0.01 '*' 0.05 '.' 0.1 ' ' 1

Residual standard error: 21.55 on 368 degrees of freedom
  (386 observations deleted due to missingness)
Multiple R-squared:  0.6355,    Adjusted R-squared:  0.5909
F-statistic: 14.26 on 45 and 368 DF,  p-value: < 2.2e-16
```

Upon analysis, our model yields an adjusted R^2 value of approximately 0.5465 or 54.65%. Interpretively, this metric tells us that about half of the variability observed in the Pokémon attack values can be explained by the predictors included in our model. Now, it might prompt curiosity as to why the model doesn't explain the full 100% of the variance, especially considering that we've incorporated a comprehensive set of data.

It's crucial to acknowledge that our dataset, albeit extensive, does not—and, indeed, cannot—encompass every conceivable factor that influences a Pokémon's attack attribute. The nature of such statistical models is to capture trends and relationships within the confines of the available data. However, in the intricate world of Pokémon, there may exist numerous hidden or latent variables that subtly or significantly impact attack stats—elements of gameplay and design choices that are outside the scope of our database.

For instance, there may be in-game evolutionary mechanics, training regimens, or player strategies that affect a Pokémon's attack strength, which our dataset doesn't capture. Additionally, the stochastic nature of Pokémon encounters and battles means that there is an inherent level of randomness and chance that no model could fully predict or explain.

6.8 Analyzing Logistic Regression

So far, we have been predicting a continuous numerical outcome, such as attack. However, what if we want to create a model that predicts a binary outcome variable, such as whether the Pokémon is a legendary type? For this, we need to use logistic regression.

In R, the only change we need to make to do logistic regression is to use a generalized linear model. Generalized linear models are a broader category of linear models that encompass the other models, including the simple linear regression models we have already seen (Nelder and Wedderburn 1972). In the GLM, residuals are no longer assumed to follow a normal distribution or have homoscedasticity. Instead, they can follow any distribution in the exponential family. To use a generalized linear model, we simply switch out the `lm()` function with the `glm()` function. This latter function allows us to specify a `family` argument that denotes the distribution of residuals. By default, it uses the "gaussian" or normal distribution. In fact, if you use `glm()` and specify that you want the Gaussian distribution, you will get the same values as `lm()`.

When doing logistic regression, we merely set the `family` argument in `glm()` to "binomial." By doing so, we signal that our outcome variable is dichotomous and that the residuals (errors) of our model will follow a binomial distribution rather than a normal distribution, which is the default for continuous data.

Let's begin by predicting legendary status based on attack, since we already know that tends to be higher for legendary Pokémon. Just as before, we can use the `summary()` function to get a comprehensive output of our model's results.

```r
# Logistic regression modeling legendary status as a function of attack
glm_1 <- glm(legendary ~ attack, family = binomial, data = pokemon)

# Summarize the logistic regression model fit
summary(glm_1)
```

```
Call:
glm(formula = legendary ~ attack, family = binomial, data = pokemon)

Coefficients:
             Estimate Std. Error z value Pr(>|z|)
(Intercept) -5.946542   0.492225 -12.081   <2e-16 ***
attack       0.037024   0.004343   8.525   <2e-16 ***
---
Signif. codes:  0 '***' 0.001 '**' 0.01 '*' 0.05 '.' 0.1 ' ' 1

(Dispersion parameter for binomial family taken to be 1)

    Null deviance: 450.90  on 799  degrees of freedom
Residual deviance: 362.63  on 798  degrees of freedom
AIC: 366.63

Number of Fisher Scoring iterations: 6
```

This output should be familiar. However, it has some important changes from general linear regression, since we are no longer predicting a continuous outcome variable on the y-axis. Since we are determining a binary value of TRUE/FALSE or 1/0, we cannot draw a straight line between the points. To compensate for this, we use an S-shape curve, more formally referred to as a "sigmoid." A sigmoid takes a value on the x-axis, in this case attack, and returns a probability between 0 and 1. We often use these values for classification rather than returning the probability. Those that have greater than 0.5 probability are given a 1, while those with less are given a 0.

In the output of our logistic regression model, as seen through the `summary()` function in R, we are presented with coefficients that are linear. This might seem counterintuitive given that we are dealing with a logistic regression, which is inherently nonlinear due to the binary nature of the response variable. What's actually happening under the hood is a transformation of the response variable that allows us to interpret the relationships in a linear framework. Instead of predicting the actual outcome y directly, logistic regression models the log-odds (or logit) of the probability of the outcome being in one category versus the other.

In the legendary status and attack logistic regression results, we see that we get a value of 0.037024 for the attack coefficient. This represents the increase in log-odds of legendary status that we can expect with a one-point increase in attack.

We also get the deviance residuals. Deviance residuals are a measure of model fit, providing insight into the difference between observed outcomes and those predicted by the model. Each residual represents a data point's contribution to the overall model deviance, which is analogous to the concept of variance in linear regression. The deviance in logistic regression is a measure of how well the model captures the pattern of the observed data. Lower deviance indicates a better fit, and examining the distribution of deviance residuals can help identify model misspecifications or particular data points that are not well explained by the model.

We also have an AIC or Akaike Information Criterion score. In this case, it is 366.63. What does this value represent? By itself, it is relatively meaningless, but we can use it to compare different models *within the same dataset*. The lower this score, the better our model fits our data. Let's see if adding speed and hp also increase the likelihood that our model can predict legendary status.

```
# Define logistic model for legendary status prediction
glm_2 <- glm(legendary ~ attack + hp + speed, family = binomial, data =
↪  pokemon)

# Display the summary of the logistic regression model
summary(glm_2)
```

```
Call:
glm(formula = legendary ~ attack + hp + speed, family = binomial,
    data = pokemon)

Coefficients:
              Estimate Std. Error z value Pr(>|z|)
(Intercept) -11.114027   1.055978 -10.525  < 2e-16 ***
attack        0.025107   0.004829   5.199 2.00e-07 ***
hp            0.033808   0.005663   5.970 2.37e-09 ***
speed         0.041433   0.006479   6.395 1.60e-10 ***
---
Signif. codes:  0 '***' 0.001 '**' 0.01 '*' 0.05 '.' 0.1 ' ' 1

(Dispersion parameter for binomial family taken to be 1)

    Null deviance: 450.90  on 799  degrees of freedom
Residual deviance: 289.46  on 796  degrees of freedom
AIC: 297.46

Number of Fisher Scoring iterations: 7
```

We get an AIC score of 297.46. This means that our new logistic model is better at explaining our data than our previous one.

We can see that determining how well a model works for logistic regression is a little different than that of linear regression. Consequently, the `check_model()` function is not very beneficial. Comparing models by AIC allows us to make sure that adding variables helps explain our phenomenon.

6.8.1 Confusion Matrices

Determining what variables increase legendary status is interesting, but we probably want to predict something more useful, such as which two Pokémon would win in a battle. To do this, we need to follow four key steps: First, we need to read in the data of battles that already have winners; second, we need to create a logistic model explaining these results; third, we need to check how well our model works on *data we already have the answer* for with a confusion matrix; and finally, if we determine that the model's results are sufficient, we can use it to predict new battles.

To set the stage for developing a predictive model, we start by importing data that encapsulates a synthetic historical record of Pokémon battles, complete with information on the combatants and the identified winners of each clash. It is crucial to remember that this battle data was generated by a custom algorithm and does not fully reflect all in-game mechanics [user feedback]. Nonetheless, this dataset, rich with labeled instances—where the "labels" refer to the known outcomes (winners) of the simulated battles—serves as the training ground for what is known in machine learning as "supervised learning."

Supervised learning is akin to an educational process where a model, much like a student, learns from examples with known outcomes. It uses the labeled synthetic historical data to discern and internalize the complex patterns and relationships that might influence the results of a battle within the dataset's simplified rules. Think of each record of a Pokémon battle as a detailed case study from this generated dataset, with the victor clearly annotated. By feeding this data into our model, we are effectively teaching it to recognize which factors—be it Pokémon type, attack strength, defense stats, or other attributes—are most predictive of battle performance within this specific, simplified dataset.

This approach contrasts with "unsupervised learning," which lacks pre-labeled outcomes. In unsupervised learning, the algorithm must sift through the data and attempt to unearth structure or patterns on its own, often grouping data points into clusters or finding patterns that define the inherent data structure. It is like a self-taught journey of discovery, where the algorithm has to make sense of the data without a guiding answer key.

Let's load our data and take a look at it.

```
# Import historical battle data from CSV file
battles <- read.csv("data/ch5_battles.csv")

# Examine battle dataset structure and contents
glimpse(battles)
```

```
Rows: 50,000
Columns: 3
$ first_pokemon  <int> 266, 702, 191, 237, 151, 657, 192, 73, 220, 302, 442, 7~
$ second_pokemon <int> 298, 701, 668, 683, 231, 752, 134, 545, 763, 31, 130, 6~
$ winner         <int> 298, 701, 668, 683, 151, 657, 134, 545, 763, 31, 130, 7~
```

In this dataset, we have three columns:

- **first_pokemon**: the id of the Pokémon that engaged in the battle (e.g., the attacker)
- **second_pokemon**: the id of the defending Pokémon
- **winner**: the id of the winning Pokémon

We need to do some additional steps to make this data useful for analysis. We begin by creating a column that indicates if the attacking Pokémon was the winner. This will give us a simple binary column that we can use to train a logistic model to predict.

```
# Create binary outcome variable for attacking Pokemon victories
battles <- battles |>
  mutate(attacker_winner = ifelse(winner == first_pokemon, 1, 0))
```

We also need to merge the dataset with our Pokémon statistics. We are going to use the `inner_join()` function to match the first and second Pokémon with the id. Since all the column names are the same, we are also going to add a suffix to show which statistics refer to the attacker and defender. We can do this by passing a vector containing our suffix to the `suffix` argument of `inner_join()`.

```
# Merge battle data with Pokemon statistics for both combatants
battles <- battles |>
  inner_join(pokemon, by = c("first_pokemon" = "id")) |>
  inner_join(pokemon, by = c("second_pokemon" = "id"), suffix =
  ↪  c(".attacker", ".defender"))
```

Now that we have our data, we can proceed to the second part of our four-part process, which is to create a logistic regression model. We want to ensure that our model includes only the variables that are integral for making predictions, specifically the Pokémon's statistical properties and the outcome of the battle, indicated by the attacking Pokémon's win status. It's important to recognize that R, the software we're using for this analysis, automatically excludes records with missing values to maintain the integrity of the model. Missing data can introduce bias and inaccuracies, so R's default behavior helps preserve the integrity of our analysis.

For the sake of simplicity and to maintain a consistent dataset, we will also exclude the secondary type attribute from our model. The rationale behind this decision is practicality—while a Pokémon's secondary type can certainly influence battle outcomes, not all Pokémon possess a secondary type. This inconsistency can lead to an uneven dataset, potentially skewing the model's results. By focusing on the primary type and other numerical statistics such as health points (hp), attack, defense, and speed, we create a more streamlined and, therefore, potentially more interpretable model.

However, it's worth noting that excluding variables, like the secondary type, is a decision that should be weighed carefully. While it simplifies the model-building process, it may also omit potentially valuable information that could enhance the model's predictive power. In the future, if a more complex model is warranted, or if we find that the primary type alone doesn't sufficiently capture the intricacies of Pokémon battles, we might consider engineering features that better encapsulate the secondary type information, such as incorporating type combinations or creating a variable that indicates the presence of a secondary type.

```
# Remove columns not needed for regression model
battles <- battles |>
  select(-(first_pokemon:winner), -name.attacker, -name.defender,
  ↪  -type_2.attacker, -type_2.defender)
```

We can now use the `glm()` function and view the `summary()` results.

```
# Calculate logistic regression model for battle outcome prediction
glm_3 <- glm(attacker_winner ~ ., data = battles, family = "binomial")

# Display logistic regression coefficients and statistics
summary(glm_3)
```

```
Call:
glm(formula = attacker_winner ~ ., family = "binomial", data = battles)

Coefficients:
                          Estimate Std. Error z value Pr(>|z|)
(Intercept)             -0.2858065  0.1017827  -2.808 0.004985 **
type_1.attackerDark      0.4519089  0.0841729   5.369 7.93e-08 ***
type_1.attackerDragon   -0.1624575  0.0845001  -1.923 0.054534 .
type_1.attackerElectric  0.1169389  0.0765964   1.527 0.126838
type_1.attackerFairy     0.0967816  0.1096821   0.882 0.377570
type_1.attackerFighting -0.1971973  0.0870824  -2.264 0.023544 *
type_1.attackerFire      0.2739566  0.0710750   3.854 0.000116 ***
type_1.attackerFlying    0.6391284  0.2312208   2.764 0.005707 **
type_1.attackerGhost    -0.2746406  0.0823489  -3.335 0.000853 ***
type_1.attackerGrass     0.1666852  0.0664952   2.507 0.012186 *
type_1.attackerGround    0.3316250  0.0832088   3.985 6.74e-05 ***
type_1.attackerIce      -0.1729882  0.0893385  -1.936 0.052828 .
type_1.attackerNormal    0.0712797  0.0621735   1.146 0.251603
type_1.attackerPoison   -0.0883359  0.0826682  -1.069 0.285268
type_1.attackerPsychic  -0.2306067  0.0731297  -3.153 0.001614 **
type_1.attackerRock     -0.0045719  0.0753446  -0.061 0.951614
type_1.attackerSteel    -0.1231729  0.0902347  -1.365 0.172244
type_1.attackerWater     0.0414849  0.0591213   0.702 0.482872
hp.attacker              0.0026810  0.0006436   4.166 3.10e-05 ***
attack.attacker          0.0090263  0.0005934  15.211  < 2e-16 ***
defense.attacker         0.0019791  0.0006023   3.286 0.001017 **
sp_attack.attacker      -0.0007300  0.0006244  -1.169 0.242361
sp_def.attacker          0.0015845  0.0006787   2.334 0.019571 *
speed.attacker           0.0659853  0.0007675  85.970  < 2e-16 ***
generation.attacker2     0.0200899  0.0478466   0.420 0.674572
generation.attacker3     0.0192546  0.0430488   0.447 0.654678
generation.attacker4     0.1258146  0.0459180   2.740 0.006144 **
generation.attacker5     0.2167114  0.0430688   5.032 4.86e-07 ***
generation.attacker6     0.1286126  0.0527115   2.440 0.014690 *
legendary.attackerTRUE  -0.1376440  0.0610864  -2.253 0.024242 *
type_1.defenderDark     -0.5556147  0.0836579  -6.642 3.10e-11 ***
type_1.defenderDragon    0.3038765  0.0853260   3.561 0.000369 ***
type_1.defenderElectric -0.0345595  0.0755645  -0.457 0.647419
type_1.defenderFairy    -0.1156302  0.1048296  -1.103 0.270014
type_1.defenderFighting  0.2084187  0.0869645   2.397 0.016548 *
type_1.defenderFire     -0.0898778  0.0711921  -1.262 0.206780
type_1.defenderFlying    0.0399357  0.2334449   0.171 0.864168
type_1.defenderGhost    -0.4703217  0.0827951  -5.681 1.34e-08 ***
type_1.defenderGrass    -0.0516604  0.0653188  -0.791 0.429005
type_1.defenderGround   -0.4084179  0.0822654  -4.965 6.88e-07 ***
type_1.defenderIce       0.1425205  0.0864522   1.649 0.099241 .
type_1.defenderNormal   -0.2931640  0.0614028  -4.774 1.80e-06 ***
type_1.defenderPoison    0.0575023  0.0815404   0.705 0.480686
type_1.defenderPsychic   0.2199364  0.0731332   3.007 0.002635 **
type_1.defenderRock      0.0594981  0.0747429   0.796 0.426011
type_1.defenderSteel    -0.3446901  0.0880627  -3.914 9.07e-05 ***
type_1.defenderWater     0.0559048  0.0583310   0.958 0.337858
hp.defender             -0.0002158  0.0006576  -0.328 0.742750
attack.defender         -0.0100178  0.0006049 -16.561  < 2e-16 ***
```

```
defense.defender          -0.0010461  0.0006038  -1.733 0.083181 .
sp_attack.defender        -0.0001535  0.0006438  -0.238 0.811581
sp_def.defender           -0.0030423  0.0006764  -4.497 6.88e-06 ***
speed.defender            -0.0641052  0.0007682 -83.448  < 2e-16 ***
generation.defender2      -0.0598167  0.0480217  -1.246 0.212905
generation.defender3      -0.1272887  0.0433587  -2.936 0.003328 **
generation.defender4      -0.1562209  0.0463189  -3.373 0.000744 ***
generation.defender5      -0.2595764  0.0436833  -5.942 2.81e-09 ***
generation.defender6      -0.1042069  0.0525577  -1.983 0.047399 *
legendary.defenderTRUE     0.1483066  0.0640595   2.315 0.020605 *
---
Signif. codes:  0 '***' 0.001 '**' 0.01 '*' 0.05 '.' 0.1 ' ' 1

(Dispersion parameter for binomial family taken to be 1)

    Null deviance: 69158  on 49999  degrees of freedom
Residual deviance: 36131  on 49941  degrees of freedom
AIC: 36249

Number of Fisher Scoring iterations: 6
```

Now that we have a model, we can go to the third step, which is to determine how well it works. There are numerous measurements we can use for this, but for now, we will utilize a confusion matrix. A confusion matrix is a simple yet illuminating table that allows us to visualize the performance of our classification model. It lays out the instances where the model's predictions align with the actual outcomes (true positives and true negatives) against the instances where the predictions and reality diverge (false positives and false negatives). By looking at the number of correct and incorrect predictions, the confusion matrix provides a snapshot of our model's predictive strength and weaknesses.

For the purposes of this preliminary assessment, we'll apply the confusion matrix to the same dataset that was used to train the model. We'll be able to see clearly how many battles were correctly predicted as wins or losses by the attacking Pokémon, providing us with an immediate sense of the model's accuracy. However, this approach does come with a caveat—by using the training data for both model building and evaluation, we run the risk of overfitting. Overfitting is akin to memorization rather than learning. The model becomes so finely tuned to the specific patterns and noise in the training dataset that it may fail to generalize to new, unseen data. This can result in a model that performs exceptionally well on the training data but poorly in practical, real-world use.

Addressing overfitting is a nuanced process that often involves techniques such as cross-validation, where the data is divided into subsets to ensure the model can perform reliably across different segments of the data. Another strategy is to hold out a portion of the dataset as a test set, never exposing the model to it during training and using it exclusively for evaluation. In Chap. 7, we will delve deeper into these strategies and explore best practices for model validation. For now, the confusion matrix will stand as our starting point for gauging the model's initial performance and sets the stage for more sophisticated validation techniques to come.

There are three steps to creating a confusion matrix. First, we create a vector containing the ground truth. This is just a fancy way of saying we need to get the column containing the "attacker_winner" column from the original data. We can use dplyr's `pull()` function to get the values into a vector.

```
# Extract actual battle outcomes for model evaluation
original_attack_winner <- battles |>
  pull(attacker_winner)
```

Next, we create a vector containing the predicted results from the model. These can be gathered using the `fitted()` function. This will return our models predictions for "attacker_winner" between 0 and 1. Since we are doing a classification task, we want to round these values, so that those above 0.5 are coded as the attacking Pokémon winning. Finally, we will convert the tibble into a vector using the `as.vector()` function.

```
# Generate binary predictions from logistic regression model
predicted_attack_winner <- glm_3 |>
  fitted() |>
  round() |>
  as.vector()
```

Our third and final step is to create a confusion matrix by using the `table()` function.

```
# Create confusion matrix to evaluate model performance
table(original_attack_winner, predicted_attack_winner)
```

```
                      predicted_attack_winner
original_attack_winner     0     1
                     0 23683  2716
                     1  2911 20690
```

Reading a confusion matrix (Predicted vs. Actual) can take some getting used to. You can read it clockwise in the following ways:

- Top-left (Predicted 0, Actual 0): Amount of times the model correctly predicted that the attacking Pokémon lost.
- Top-right (Predicted 1, Actual 0): Amount of times the model predicted the attacking Pokémon won when it did not.
- Bottom-right (Predicted 1, Actual 1): Amount of times the model correctly predicted that the attacking Pokémon won.
- Bottom-left (Predicted 0, Actual 1): Amount of times the model predicted the attacking Pokémon lost when it did not.

We can see from the above results that we have a fairly good model that predicts over 90 percent of our battles correctly. However, it bears repeating that this model may only be good at working with our training data rather than new data. We will look at how to split our data up in Chap. 7, but for now, we will go ahead and use this model to predict winners.

6.8.2 Predicting Test Data

Now that we know we have a good model, we are finally ready to predict the results for new battles. First, we will load our CSV containing the battles needing to be predicted and observe it.

```
# Import new battle data requiring outcome predictions
battles_needing_prediction <-
↪  read_csv("data/ch5_battles_needing_prediction.csv")

# Examine structure of prediction dataset
glimpse(battles_needing_prediction)
```

```
Rows: 10,000
Columns: 2
$ first_pokemon  <dbl> 129, 660, 706, 195, 27, 126, 436, 121, 169, 302, 39, 11~
$ second_pokemon <dbl> 117, 211, 115, 618, 656, 222, 207, 36, 636, 688, 380, 2~
```

We see that our columns are the same as our training data minus the column that states the winner. When using a model for prediction, all the columns should have the properties used for prediction. Thus, we need to combine our Pokémon statistics to this dataset as we did when we trained the data.

```
# Merge new battles with Pokemon statistics for prediction
battles_needing_prediction <- battles_needing_prediction |>
  inner_join(pokemon, by = c("first_pokemon" = "id")) |>
  inner_join(pokemon, by = c("second_pokemon" = "id"), suffix =
  ↪  c(".attacker", ".defender"))
```

And, just like before, we also need to clean up the columns we no longer need.

```
# Filter prediction dataset to match training data structure
battles_needing_prediction <- battles_needing_prediction |>
  select(-first_pokemon, -second_pokemon, -name.attacker, -name.defender,
  ↪  -type_2.attacker, -type_2.defender)
```

Finally, we can predict the results using the `predict()` function, which takes the model as the first argument and the data needing to be predicted as the second. We will also set the `type` argument to "response," which sets the scale of the predictions not on the scale of the linear predictions but the scale of the response variable. In this case, our response variable will be given as log-odds. Finally, we will round these up if they are 0.5 or higher and down if they are not before printing the results.

```
# Generate battle outcome predictions using trained model
battles_needing_prediction$attacker_winner <- predict(glm_3,
↪  battles_needing_prediction, type = "response") |>
  round()

# Examine final prediction results
glimpse(battles_needing_prediction)
```

```
Rows: 10,000
Columns: 19
$ type_1.attacker     <chr> "Water", "Grass", "Electric", "Electric", "Normal"~
$ hp.attacker         <dbl> 80, 74, 79, 70, 40, 30, 44, 105, 80, 40, 61, 59, 6~
$ attack.attacker     <dbl> 92, 94, 105, 55, 60, 40, 58, 130, 82, 30, 72, 63, ~
$ defense.attacker    <dbl> 65, 131, 70, 55, 30, 70, 44, 120, 100, 30, 57, 80,~
$ sp_attack.attacker  <dbl> 65, 54, 145, 80, 31, 70, 58, 45, 83, 55, 55, 65, 5~
$ sp_def.attacker     <dbl> 80, 116, 80, 60, 31, 25, 44, 45, 100, 30, 55, 80, ~
$ speed.attacker      <dbl> 68, 20, 101, 45, 70, 60, 61, 40, 80, 85, 65, 58, 8~
$ generation.attacker <fct> 1, 5, 5, 2, 1, 1, 4, 1, 2, 3, 1, 1, 2, 3, 4, 3, 2,~
$ legendary.attacker  <lgl> FALSE, FALSE, TRUE, FALSE, FALSE, FALSE, FALSE, FA~
$ type_1.defender     <chr> "Normal", "Water", "Fighting", "Grass", "Water", "~
$ hp.defender         <dbl> 90, 95, 50, 75, 165, 100, 30, 70, 45, 95, 45, 50, ~
$ attack.defender     <dbl> 55, 85, 120, 86, 75, 70, 30, 62, 30, 110, 95, 35, ~
$ defense.defender    <dbl> 75, 85, 53, 67, 80, 70, 30, 67, 50, 95, 50, 55, 76~
$ sp_attack.defender  <dbl> 60, 65, 35, 106, 40, 65, 30, 55, 55, 40, 40, 25, 8~
$ sp_def.defender     <dbl> 75, 65, 110, 67, 45, 65, 30, 55, 65, 95, 50, 25, 8~
$ speed.defender      <dbl> 30, 35, 87, 60, 65, 45, 30, 56, 45, 55, 75, 15, 10~
$ generation.defender <fct> 1, 2, 1, 5, 5, 2, 2, 1, 5, 5, 3, 3, 6, 6, 4, 1, 5,~
$ legendary.defender  <lgl> FALSE, FALSE, FALSE, FALSE, FALSE, FALSE, FALSE, F~
$ attacker_winner     <dbl> 1, 0, 1, 0, 0, 1, 1, 1, 1, 1, 0, 1, 0, 1, 1, 1, 0,~
```

6.9 Conclusion

This chapter was relatively complex, so don't worry if you didn't understand everything immediately. Even those who have been working with these models for a while find some of the details perplexing. However, prediction and inference through regression are powerful techniques worth learning. We saw how we could use them to make sense of our Pokémon dataset, particularly the synthetically generated battle results. Specifically, we analyzed how regression, both linear and logistic, can be leveraged to make sense of the myriad factors that influence battle outcomes within this simplified dataset. From examining the fundamental characteristics like attack and defense (derived from the actual games) to more complex features such as type and generation, our analysis has shed light on the statistical underpinnings of battle dynamics as represented in this specific dataset, offering insights into a complex domain through a quantitative lens.

We delved deeply into the statistical method of regression, starting with the simpler linear regression to predict continuous outcomes like the number of books sold or a student's grade. We then transitioned to logistic regression, which allowed us to predict binary outcomes, such as the likelihood of a Pokémon being legendary or the chances of a Pokémon winning a battle. Through these exercises, we not only learned how to fit and interpret regression models but also how to critically evaluate their performance using tools like confusion matrices and the concept of overfitting.

The techniques we've covered here lay the groundwork for more advanced statistical modeling. While linear and logistic regressions are powerful tools, they are just the tip of the iceberg in the expansive sea of machine learning methodologies. In Chap. 7, we'll move beyond these foundational concepts and delve into the realm of advanced machine learning. We'll explore how algorithms can learn even more complex patterns from data, make more accurate predictions, and uncover deeper insights.

Chapter 7
Tidy Networks

Abstract This chapter delves into the application of network analysis within the R ecosystem, utilizing the "tidygraph" package to analyze the Early Race Film Database. By leveraging the tidy principles and integrating them with network-specific analysis, it uncovers the intricate web of interactions among films, actors, directors, and production companies. The chapter demonstrates how to import and wrangle network data, calculate centrality measures, visualize connections, and detect communities, revealing the latent structures of cultural networks. It overviews visualization techniques, facilitated by the "ggraph" package, and how they translate complex relational data into insightful diagrams.

Keywords Network theory · "Tidygraph" · Centrality measures · Race films · Network visualization · Community detection · "Ggraph" · Interconnectivity · Cultural networks

7.1 Introduction

In the realm of data analysis, one fundamental assumption that underpins many statistical techniques is the independence of variables. Classical methods such as linear regression, for instance, are predicated on this and require that the variables within a model not exhibit multicollinearity. However, the complex nature of human behavior and societal interactions often defies this requirement. In the humanities and social sciences, for instance, the data we encounter is not neatly compartmentalized; rather, it is richly interwoven with variables influencing each other in a web of interdependence.

This interdependence is not merely a challenge to be overcome but a phenomenon of profound interest, reflecting the intricate tapestry of human relationships and societal structures. How do we capture this interdependence in our analyses? How do we sift through the noise to identify variables within cultural phenomena that wield the most influence? How do we determine the strength and significance of ties that bind objects and people together? Perhaps most crucially, how do we uncover the underlying structures within cultural networks—the clusters

N. Siddiqui, *Cultural Analytics in R: A Tidy Approach*, Use R!,
https://doi.org/10.1007/978-3-031-96618-7_7

and communities that emerge organically from the myriad interactions? Network theory offers a powerful lens through which to answer these questions (*Statistical Analysis of Network Data with R* 2020). It is built upon a simple principle that is both intuitive and profound: Entities that interact more often with each other influence each other more. This might seem like an obvious observation, yet it has far-reaching implications for understanding complex systems. Through network theory, we can map dynamics within a system, whether these are people interacting on social media, scientists citing each other's work, or words co-occurring in a text corpus.

In this chapter, we will venture into the intricate world of networks to uncover the hidden patterns and connections within cultural data. Within the R ecosystem, several packages are dedicated to network analysis, each offering a unique set of tools and approaches. Among these, "igraph" is a particularly versatile and widely adopted package. Its comprehensive suite of functions can visualize relational data and perform intricate calculations (Csardi and Nepusz 2006). From simple centrality measures to more complex algorithms for detecting community structures, "igraph" is a beneficial tool for network scientists seeking to quantify and interpret connections among data points. Rather than working directly with "igraph," however, we will examine the capabilities of the "tidygraph" package by Thomas Lin Pedersen (Pedersen 2023). This package serves as a wrapper around "igraph," marrying it with the principles of the "tidyverse." Thus, it allows researchers already familiar with "dplyr" to apply the same intuitive and powerful data-wrangling techniques to networks while benefiting from the statistical capabilities of "igraph" and other similar packages.

Although statistical measures derived from networks are informative and can reveal much about the structure and importance of the relationships within, they are often abstract and may not immediately convey the richness of the underlying data to a broader audience. Visualization is a crucial step in making the insights gleaned from network analysis more accessible and interpretive. To this end, Pedersen developed a companion visualization package called "ggraph," which is rooted in the grammar of graphics (Pedersen 2022). Paired with "tidygraph," it provides a formidable toolkit for the analysis and communication of network data within a tidy framework.

Throughout our journey, we will examine the Early Race Film Database to delve into the practical applications of network analysis (Cifor et al. 2018b). Race films, a genre largely produced between the 1910s and the 1950s, were created by and for African Americans, offering perspectives largely absent from "mainstream" productions of the era. Yet, scholarship on them is scant. Underlying this lack of scholarship are disputes about what exactly constitutes race films. As the curators of the database note, "race film was never a rigid category; it was flexible, capacious, and contested" (Cifor et al. 2018a). Nonetheless, while keeping this in mind, exploring this database through network analysis can help us understand the relationships between films, actors, directors, and production companies. By looking at variables such as shared actors or directors, we can identify key figures in this cinematic movement. Luckily, the authors of the database have done a lot of preliminary research and have excellent tutorials on how to work with this data using mostly graphical tools. Our goal will be to recreate and expand this analysis in R (Posner and Lincoln 2016).

By the end of this chapter, you should have a strong foundation in network data and how it can answer research questions in cultural analytics. You will know how to calculate and interpret key network statistics—such as measures of centrality and community detection algorithms—that can unveil the most influential nodes and the structure of clusters within your network. Furthermore, you will understand how to visualize networks using the "ggraph" package, translating intricate webs of relationships into insightful diagrams that can communicate your findings effectively. The goal is to prepare you to tackle other research questions in cultural analytics using the tools and approaches discussed here.

7.2 Importing Early African American Film Data

We will begin by importing our data using the `read_csv()` function and subsequently examining it using `glimpse()`. The dataset codes missing data as blank values; as a result, we need to convert these to R's NA type by passing a blank to the `read_csv()` function's `na` argument.

```
# Load the dataset by reading the CSV and treat blanks as NAs
films <- read_csv("data/ch6_race_films.csv", na = "")

# Output a concise overview of the dataset's structure and contents
glimpse(films)
```

```
Rows: 759
Columns: 10
$ last_name         <chr> "Abbott", "Abbott", "Abrahams", "Abrams", "Agusta", "~
$ first_name        <chr> "Robert", "Mrs. Robert", "Edward R.", "Ed", "Howard",~
$ full_name         <chr> "Robert Abbott", "Mrs. Robert Abbott", "Edward R. Abr~
$ aka               <chr> NA, NA, NA, "Dick Abrams", NA, NA, NA, NA, NA, NA, NA~
$ dob               <chr> NA, NA, NA, NA, NA, NA, NA, NA, NA, NA, NA, NA, NA, N~
$ dod               <chr> NA, NA, NA, NA, NA, NA, NA, NA, NA, NA, NA, NA, NA, N~
$ films_appeared    <chr> "The Millionaire", "The Millionaire", "The Sport of t~
$ films_directed    <chr> NA, NA, NA, NA, NA, NA, NA, NA, NA, NA, NA, NA, NA, N~
$ films_produced    <chr> NA, NA, NA, NA, NA, NA, NA, NA, NA, NA, NA, NA, NA, N~
$ films_camerawork <chr> NA, NA, NA, NA, NA, NA, NA, NA, NA, NA, NA, NA, NA, N~
```

We see that our table has 759 rows with the following 10 columns:

last_name: The last name of the individual involved in the film.
first_name: The first name of the individual involved in the film.
full_name: The full name of the individual involved in the film.
aka: Some individuals used multiple names throughout their career or used names that were not associated with their birth names.

dob: The date of birth of the individual listed.

films_appeared: The names of the films an individual acted in separated by commas.

films_directed: The names of the films an individual directed separated by commas.

films_produced: The names of the films an individual produced separated by commas.

films_camerawork: The names of the films an individual did some sort of camera work in separated by commas.

7.3 Preliminary Data Exploration of Early African American Film

Since a significant portion of this chapter is devoted to unraveling the complexities of network visualization, our exploration into alternative plotting techniques will be somewhat brief. To fully appreciate the significance of these visualizations, let's contextualize the environment from which these race films emerged. The early twentieth-century race film industry operated under financial constraints that were markedly more stringent than those faced by its mainstream counterparts. This economic disparity resulted in a film industry that had to be exceptionally innovative with its resources. It was not just common but expected for individuals involved in the creation of race films to possess a diverse array of talents; their roles often spanned directing, acting, and camera work within the same production. This multifaceted involvement was more than a mere response to budgetary limitations—it was a profound demonstration of the resourcefulness, resilience, and boundless creativity that characterized the African American film community during this era (Diawara 1993; Cripps 1979; Gaines 2001).

When examining our dataset, we can see that some individuals were key to the industry and contributed to multiple film projects. Such contributions are indicated in our data by a comma-separated list of film titles associated with each name. For instance, Howard August's name appears alongside both *Smiling Hate* and *Children of Fate*. How this data is currently structured, however, presents a challenge for conducting straightforward quantitative analyses. Simple yet fundamental questions—such as the tallying of film appearances per actor or identifying the films with the most diverse casts—are not easy to answer due to the nonstandardized format of the data. Consequently, we will often need to separate the films into individual observations to simplify the analytical process.

Let's begin by looking at the ten most common performers in our data with a bar graph. We will use the `separate_rows()` function on the "films_appeared" column to handle cases where a performer appeared in multiple films. However, we will also omit `NA` values. This is due to the fact that our data also has individuals who did not perform on-screen but participated in some other capacity, such as scriptwriting or directing. For now, we will try to spotlight only those in front of the camera for this particular exploration, but later, you may want to examine the full dataset.

```
# Count the number of films each performer appeared in and visualize the
↪  top ten in a column chart
films |>
  separate_rows(films_appeared, sep = ",") |>
  filter(!is.na(films_appeared)) |>
  count(full_name, sort = TRUE) |>
  top_n(10) |>
  ggplot(aes(x = full_name, y = n)) +
  geom_col() +
  coord_flip()
```

Figure 7.1 shows that Lawrence Chenault emerges as a particularly prolific figure, having graced the screen in 22 films. Close on his heels are performers Will Starks, Sam Robinson, and Samuel Jacks. These numbers might initially suggest a concentrated pool of talent frequently tapped for multiple projects within the race film industry. But do these figures truly signify an exceptional level of involvement, or were multiple film appearances the norm during this era?

To contextualize the data and gain a better understanding of the broader patterns of participation among performers, we can construct a histogram to visualize the distribution of film appearances across all actors. A histogram that skews heavily toward fewer appearances would suggest that many actors had brief forays in the race film scene, possibly indicating a vast network of transient talent alongside a smaller core of regulars. Conversely, a more uniform distribution or one that leans toward higher appearance counts could indicate a robust, tightly knit community of performers with regular engagements.

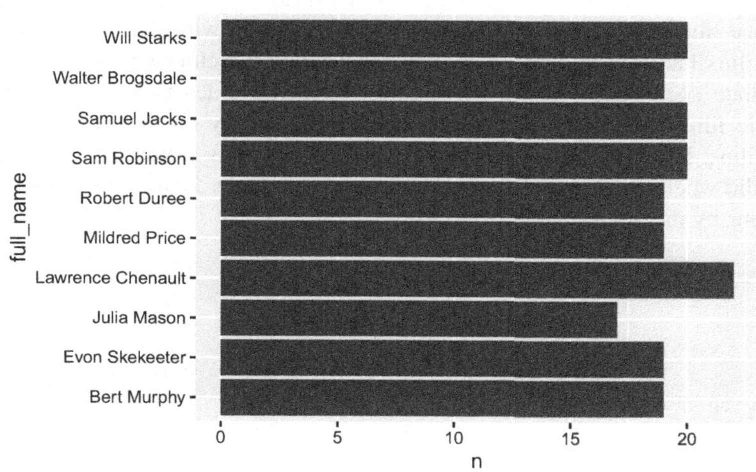

Fig. 7.1 Top 10 performers by number of film appearances in the Early Race Film Database

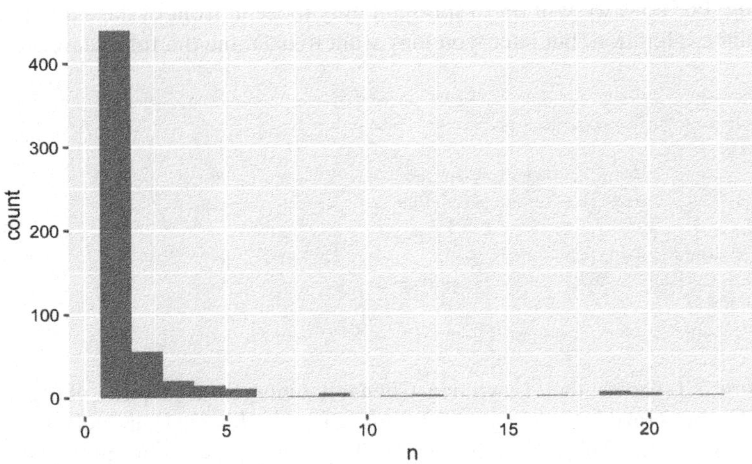

Fig. 7.2 Distribution of the number of film appearances per performer

```
# Create a histogram to visualize the distribution of film appearances
↪ among performers but with a larger bin size
films |>
  separate_rows(films_appeared, sep = ",") |>
  filter(!is.na(films_appeared)) |>
  count(full_name) |>
  ggplot(aes(x = n)) +
    geom_histogram(bins = 20)
```

Figure 7.2 shows that most individuals only appeared in a single film. This can lend us insight into how race films functioned. For many performers, acting in a race film was probably a novel experience. Along with the fact that most of these films had tiny budgets, it is likely that many directors reached out to their immediate social circle to fill small roles. Even those actors who did appear in multiple films presumably did not make a living purely on their appearances in these films. However, we may be interested in what films defied these norms. Just as we did when we plotted the top performers, we can use a bar graph to see the top five films by the number of performers.

```
# Count the number of performers in each film and visualize the top five
↪ in a column chart
films |>
  separate_rows(films_appeared, sep = ",") |>
  filter(!is.na(films_appeared)) |>
  count(films_appeared, sort = TRUE)   |>
  top_n(5) |>
  ggplot(aes(x = films_appeared, y = n)) +
    geom_col() +
    coord_flip()
```

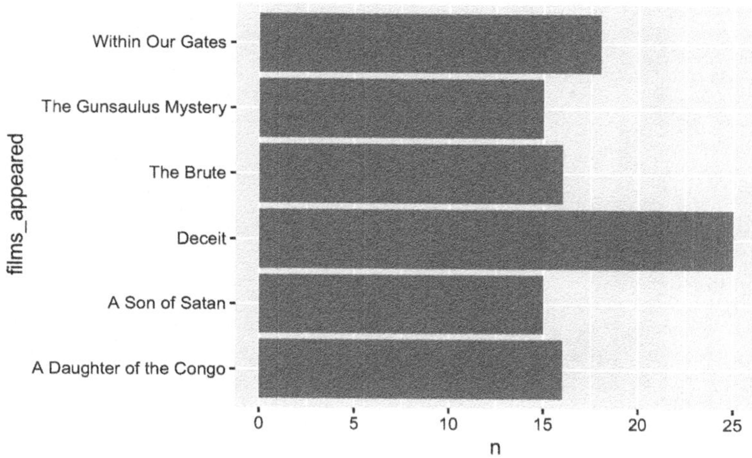

Fig. 7.3 Top 5 films with the largest number of performers listed

As shown in Fig. 7.3, the 1923 Oscar Micheaux film *Deceit* stands out for its expansive cast, numbering 25 performers—a significant figure, especially in comparison to its contemporaries. Trailing closely behind *Deceit* was another of Micheaux's acclaimed works, *Within Our Gates*, which only further underscores his prominence in the industry. Due to his acumen, he was able to marshal resources with an efficacy that was unparalleled at the time (Bowser et al. 2016; Bowser and Spence 2000). This financial backing, combined with his visionary leadership, resulted in a body of work that was both pioneering and prolific, with his hands in the production of all the top films we have identified.

The exceptional cast size of Oscar Micheaux's films such as *Deceit* and *Within Our Gates* might lead us to wonder whether these productions were outliers in the broader landscape of race films. With Micheaux's films featuring a high number of performers, it is natural to question if this was a common practice or if Micheaux's works were indeed singular in their scale. To investigate this further and test our hypothesis regarding the anomaly of Micheaux's casting practices, we can employ a histogram.

```
# Create a histogram to visualize the distribution of film appearances
↪  across all films
films |>
  separate_rows(films_appeared, sep = ",") |>
  filter(!is.na(films_appeared)) |>
  count(films_appeared) |>
  ggplot(aes(x = n)) +
    geom_histogram(bins = 25)
```

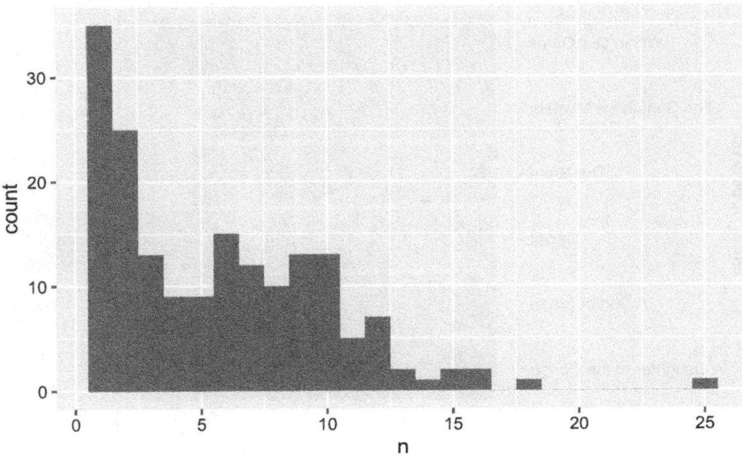

Fig. 7.4 Distribution of the number of performers per film

Figure 7.4 confirms our hunch was correct. A vast majority of these films typically cast a relatively small number of performers, with the common range being between one and ten individuals. This lean approach to casting had implications for the production process. It meant that those in the film had to be judicious in their storytelling so as to convey powerful narratives within the confines of smaller ensembles. Moreover, the small cast size may have fostered a sense of intimacy and community among the actors and crew, which could have translated into the chemistry evident on screen. For audiences of the time, seeing familiar faces also added an extra layer of engagement and loyalty to the burgeoning genre. To examine these relationships more closely and understand the broader impact of casting choices, we will delve into network analysis. By mapping the connections between performers and the films in which they appeared, our goal will be to uncover patterns that will allow us to identify other key players within the community.

7.4 Network Theory Basics

At the heart of network theory lies a basic yet powerful structure composed of two fundamental components: nodes and edges. Understanding the mechanics and relationships between these two elements is crucial to harnessing the full potential of network analysis.

Nodes, or vertices as they are sometimes called, are the primary building blocks of a network. They represent the discrete entities that we wish to analyze and understand. In a social network, for example, individuals are the nodes. In a transportation network, cities or intersections might serve as nodes. The choice of what constitutes a node is determined by the focus of the investigation, and it is this choice that sets the stage for the insights that the network can reveal. For instance, in the case of our study of race films, the nodes are twofold: the performers and the films themselves.

Nodes are typically represented visually as circles, a matter of tradition rather than necessity. The node's meaning comes from what it symbolizes—the underlying data point or entity. Circles offer simplicity and clarity, but any shape or symbol that clearly distinguishes one node from another can serve the same purpose. When plotting these circles within a network graph, they are assigned a position in a two-dimensional space where they can be connected by edges. It is important to note that, in most cases, this placement has no inherent meaning and is only used to facilitate an understanding of the network's underlying structure. In other words, the location of the node is strategic, aiming to minimize visual clutter, like overlapping lines or crowded clusters, which can obscure important connections and patterns.

The diversity of nodes within a network can give rise to different types of graphs, each with its unique characteristics and analytical approaches. If we deal with only one type of node, we are working with what is known as a unimodal or unipartite graph. Unimodal graphs have been extensively studied and are well-understood within the fields of mathematics, computer science, and network theory, making them a solid foundation for many statistical analyses. However, the complexity of real-world interactions often requires us to move beyond unimodal graphs. When our study involves more than one type of node, we enter the realm of multimodal or multipartite graphs. These graphs acknowledge the multifaceted nature of networks by allowing for different node categories. For example, a bimodal graph contains two types of nodes, such as the performers and films in our case, creating two distinct but interrelated sets of vertices. Similarly, a trimodal graph would involve three types of nodes, and so on, with each additional mode adding a layer of complexity to the network.

Edges, or links, are the conduits through which nodes interact, forming the relational architecture of the network. They can be either undirected, signifying a bidirectional or mutual relationship, or directed, indicating a one-way relationship where the directionality of the interaction is crucial to understanding the nature of the connection. For instance, an undirected edge in a friendship network implies mutual recognition of the friendship, while a directed edge in a citation network indicates the flow of influence from the cited work to the citing work. The visual representation of directed edges typically involves arrows that point from the source node to the target node, making the direction of the relationship explicit, enhancing the graph's interpretability.

Understanding the types of edges is vital for accurate network analysis, as it informs the interpretation of the connections between nodes. For example, in our bimodal graph of performers and films, we could represent a performer's participation in a film with an undirected edge, suggesting a reciprocal relationship: The film features the performer, and the performer is part of the film's cast. However, if we wanted to analyze the influence of a particular film on a performer's subsequent roles, we might opt for directed edges that point from the film to the performer, indicating the direction of influence. This distinction between undirected and directed edges becomes even more pronounced when we consider the dynamics of influence, power, or information flow within the network. Directed edges allow us to map out complex hierarchies, dependencies, and asymmetric relationships that often exist within social structures and technological networks.

Thus, by combining nodes and edges to construct and analyze our data, we seek to capture the interplay between films and performers, not merely plotting static points and connecting lines. We are, in essence, weaving a narrative that reveals the multifaceted connections between the artistic endeavors of the film industry and the individual careers of the actors involved.

7.5 Understanding Tidygraph

The tidy data paradigm emphasizes the importance of structuring datasets in a way that each variable forms a column and each observation forms a row. While this approach brings clarity and efficiency to many data analysis scenarios, it encounters a unique challenge when faced with the inherent duality of network data. Networks, by their very nature, consist of entities (nodes) and the connections between them (edges), each of which could be considered observations in their own right. Trying to force them into one table would not only be impractical but would also obscure the relationships that give a network its character and analytical power. This is where the complexity of network data clashes with the tidy data principles (Pedersen 2017).

"Tidygraph" tries to solve this by integrating the principles of tidy data with the requirements of network analysis. It offers a cohesive framework that retains the separate identities of nodes and edges while keeping them within a single, navigable object. This approach recognizes that while nodes and edges are separate entities, they are also intrinsically linked. Put simply, although we will work with one "tidygraph" object, we are actually working with two tibbles.

Deftly navigating within this dual structure requires a mechanism to switch the analytic focus between the two components—nodes and edges—depending on the specific aspect of the network under examination. "Tidygraph" introduces the `activate()` function as an elegant solution to this challenge. This function acts like a switch, enabling the analyst to toggle between node-centric and edge-centric views. This pivotal feature allows the user to direct the full power of the "dplyr" toolkit to the appropriate part of the network graph as needed.

By simply passing the argument "nodes" or "edges" to `activate()`, we instruct R on which subset of the "tidygraph" object we intend to work with. Once activated, a particular subset can be manipulated with "dplyr" verbs to perform a wide array of data operations. For instance, we can leverage 'dplyr's `filter()` function to hone in on specific nodes or edges, use `mutate()` to create or transform variables within our node or edge data, and employ `summarize()` to distill our data down to key metrics or insights.

Let's go ahead and install and load "tidygraph" to see how all of this works.

```
```{r}
# Install the "tidygraph" package from CRAN
install.packages("tidygraph")
# Load the "tidygraph" package into the R session
library(tidygraph)
```
```

7.5.1 Creating an Edge List

Having successfully installed our network analysis software, we can embark on the actual construction of our network. Typically, in network analysis, the initial step is to define all the nodes. This is a logical starting point, as it allows us to lay down the entities that we will later interconnect through various relationships. Conceptually, it's akin to mapping out the "cast of characters" before delineating the web of interactions that links them. And while "tidygraph" does afford us the ability to manually input each node, it offers a more efficient route—by constructing nodes implicitly through a tibble of edges.

The dataset on films and performers necessitates a few preparatory steps to create the edge tibble. Our primary purpose will be on those individuals who appeared on screen, prompting us to select only the columns pertaining to full names and their respective film appearances—this distillation of data helps maintain a clear focus on our subjects of interest. Next, we address the intricacy of performers' filmographies, where a single cell might list multiple film credits. To navigate this, we will do a row separation, ensuring that each film credit occupies its own row and thus forms a distinct edge in our network. Finally, we will remove NA values, which represent individuals in our dataset who did not appear on-screen.

After this preparatory step, we can use the `tbl_graph()` function to create the network. When we feed it edge data, it recognizes the columns that represent the "source" and "target" nodes—typically the first two columns in our dataset. This intelligent design simplifies the network creation process, allowing us to focus on the relationships we aim to study without getting bogged down by the minutiae of node entry.

```
# Filter the dataset to include only relevant columns and separate rows
↪  for films
films_filtered <- films |>
  select(full_name, films_appeared) |>
  separate_rows(films_appeared, sep = ",") |>
  filter(!is.na(films_appeared))
```

```
# Create a tidygraph object from the filtered edge data
films_tidygraph <- tbl_graph(edges = films_filtered)
```

Now that we have created our “tidygraph” object, let’s take a look at it.

```
# Examine tidygraph object
films_tidygraph
```

```
# A tbl_graph: 731 nodes and 990 edges
#
# A directed acyclic multigraph with 48 components
#
# Node Data: 731 x 1 (active)
   name
   <chr>
 1 Robert Abbott
 2 The Millionaire
 3 Mrs. Robert Abbott
 4 Edward R. Abrahams
 5 The Sport of the Gods
 6 Ed Abrams
 7 The Gunsaulus Mystery
 8 Howard Agusta
 9 Smiling Hate
10 Children of Fate
# i 721 more rows
#
# Edge Data: 990 x 2
   from    to
  <int> <int>
1     1     2
2     3     2
3     4     5
# i 987 more rows
```

With 731 nodes and 990 edges identified, we are provided with a quantifiable snapshot of our network’s scale and complexity. As we can see, although “tidygraph” constructed our network using only the specified edge data, it proactively assembled not just one but two structured tibbles within the resulting object. The first of these tibbles is a comprehensive list of nodes; in our case, the node tibble encompasses data from both films and performers. This clearly represents the bimodal nature of our network, where two distinct types of entities interact. While “tidygraph” initially treats all nodes with an egalitarian approach, our bimodal network would benefit from distinguishing between these two types of nodes, a process we will explore later in this chapter.

The second tibble generated by "tidygraph" consists of edges. It becomes apparent that our input has been adeptly transformed to align with network analysis conventions. Where we once had columns for performers and "films_appeared," we now see the nomenclature has shifted to "from" and "to." These labels are not merely cosmetic changes but a crucial adaptation for the graph structure, embodying the directionality of the relational data. By default, "tidygraph" creates directed edges, which can be problematic if you later run statistical measures on your graph. Luckily, it is easy to specify that you are interested in an undirected graph by setting the `directed` argument to `FALSE`. For now, we will stick with a directed network. Thus, each edge denotes a directional flow from the node labeled "from" (the performer) to the node labeled "to" (the film). However, it's important to note that not all networks possess this inherent directionality.

The transformation within the edges tibble extends beyond the column names. The values they contain have also been redefined to correspond to unique identifiers or row IDs of the nodes. This re-indexing process enables "tidygraph" to streamline the association between nodes and their edges, creating a clear and efficient mapping that is essential for network computation and visualization.

7.5.2 *The Activate Function*

When working within the "tidygraph" framework, selecting a subset of the data—such as the first ten nodes—presents a unique challenge due to the presence of two interconnected tibbles within the same graph object. The question arises: When we invoke the `slice()` function to view a segment of our data, how do we specify whether we're referring to the nodes or the edges? This is where "tidygraph's" innovative `activate()` function comes into play. The `activate()` function is essentially a context setter. It defines the scope of our data manipulation efforts by toggling the active status between nodes and edges. By passing the argument "nodes" to `activate()`, we instruct "tidygraph" to focus on the node tibble for any subsequent operations within the chain of commands. This means that when we then use the `slice()` function, "tidygraph" understands that we are referring to the first ten rows of the node tibble rather than the edges.

```
# Activate the nodes part of the tidygraph object and slice the first ten
↪  rows
films_tidygraph |>
  activate(nodes) |>
  slice(1:10)
```

```
# A tbl_graph: 10 nodes and 6 edges
#
# A rooted forest with 4 trees
#
```

```
# Node Data: 10 x 1 (active)
   name
   <chr>
 1 Robert Abbott
 2 The Millionaire
 3 Mrs. Robert Abbott
 4 Edward R. Abrahams
 5 The Sport of the Gods
 6 Ed Abrams
 7 The Gunsaulus Mystery
 8 Howard Agusta
 9 Smiling Hate
10 Children of Fate
#
# Edge Data: 6 x 2
   from    to
  <int> <int>
1     1     2
2     3     2
3     4     5
# i 3 more rows
```

As you see, we only get the first ten nodes and their corresponding edges. Technically, we did not need to specify that we wanted to activate nodes as that is the default activation. However, I have made it explicit above and will continue to do so throughout the rest of the chapter for clarity.

7.5.3 Essential Network Visualization with Tidygraph

The `plot()` function in R serves as a pivotal tool for visualizing graphs, transforming abstract network structures into tangible illustrations. When we plot a network, the nodes are positioned within a virtual space, and their placement is determined by a layout algorithm designed to spatially separate nodes in a way that the edges between them can be easily distinguished. As mentioned earlier, the placement of nodes in this visual representation doesn't have intrinsic meaning in terms of the network's properties—the spatial arrangement is purely for visualization purposes. However, this arbitrary placement introduces challenges, especially when dealing with networks that contain a large number of nodes. In such cases, the potential for nodes to overlap on the plot increases significantly, obstructing the viewer's ability to discern the individual nodes and the connections between them. Overlapping nodes create visual clutter and compromise the legibility of the network, obscuring the very relationships we seek to understand. To circumvent this issue and ensure a clear and comprehensible visualization, it may be necessary to limit the number of nodes displayed, as we have done by focusing on the first ten nodes. This selective approach allows us to maintain an unobstructed view of these nodes and their immediate connections, providing us with an accessible and informative snapshot of the network.

Given that the layout algorithm functions with a degree of randomness, the positioning of nodes can vary with each execution of the `plot()` function. This randomness, while valuable for exploring different visual arrangements, poses a challenge for consistency and reproducibility. To address this, we can employ the `set.seed()` function in R, which initializes the random number generator to a fixed point, effectively "locking in" the layout of nodes for a given seed value. By doing this, we guarantee that each time we run the plot function with the same seed, we will obtain an identical arrangement of nodes. The choice of the seed number is at the discretion of the analyst and does not affect the layout's quality. For instance, we will choose the seed value of "123." Should a different arrangement be desired, one can simply choose a new seed value or omit the seed setting for a fresh, randomized layout with each plot.

With all this in mind, let's look at the plot() function to visualize the first ten nodes.

```
# Set random seed for consistent node layout in the plot and visualize
↪  the first ten nodes
set.seed(123)
  films_tidygraph |>
  activate(nodes) |>
  slice(1:10) |>
  plot()
```

Figure 7.5 demonstrates that the built-in `plot()` function is good for quick analysis but is limited in capacity. We will soon see how to use "ggraph" for more advanced plots, which adds geometric objects specific to networks into a grammar of graphics framework. Still, even in this rudimentary graph, we can see some issues with our graph: Our data involves two distinct types of nodes (films and performers).

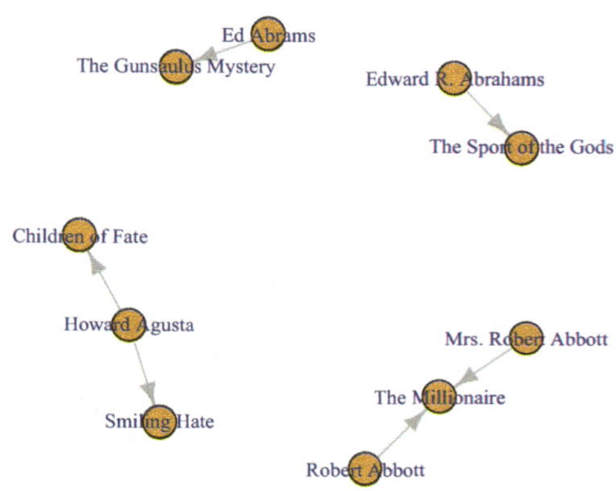

Fig. 7.5 Basic R plot showing the first ten nodes of the film-performer network

This distinction is crucial for accurate interpretation, as it helps viewers immediately recognize the different roles that nodes play within the network.

To enhance the utility of our network graph and prepare for a more advanced visualization, we need to enrich our node data with additional attributes that capture the essence of each node beyond just its name. We achieve this by adding a category column to our node dataset, indicating the type of node—film or a performer. This is accomplished using the `node_is_source()` function from the "tidygraph" package. By applying this function, we can assess whether a node appears in the "from" column of our edge data—typically indicative of a source node in directed networks—and assign the appropriate category based on this criterion. To enhance the utility of our network graph and prepare for a more advanced visualization, we need to enrich our node data with additional attributes that capture the essence of each node beyond just its name. We achieve this by adding a "category" column to our node dataset, which will indicate the type of node—film or a performer. This is accomplished using the `node_is_source()` function from the "tidygraph" package. By applying this function, we can assess whether a node appears in the "from" column of our edge data—typically indicative of a source node in directed networks—and assign the appropriate category based on this criterion.

```
# Add a category column to the node tibble, classifying nodes as either
↪  performer or film
films_tidygraph <- films_tidygraph |>
  activate(nodes) |>
  mutate(category = ifelse(node_is_source(), "performer", "film"))
```

Let's take a look at our results:

```
# Examine tidygraph object
films_tidygraph
```

```
# A tbl_graph: 731 nodes and 990 edges
#
# A directed acyclic multigraph with 48 components
#
# Node Data: 731 x 2 (active)
   name                    category
   <chr>                   <chr>
 1 Robert Abbott           performer
 2 The Millionaire         film
 3 Mrs. Robert Abbott      performer
 4 Edward R. Abrahams      performer
 5 The Sport of the Gods   film
 6 Ed Abrams               performer
 7 The Gunsaulus Mystery   film
 8 Howard Agusta           performer
 9 Smiling Hate            film
10 Children of Fate        film
```

```
# i 721 more rows
#
# Edge Data: 990 x 2
   from    to
  <int> <int>
1     1     2
2     3     2
3     4     5
# i 987 more rows
```

With the introduction of a category column that classifies each node as either a performer or a film, our network data gains an additional dimension that is ripe for exploration. By default, "tidygraph" calculates centrality measures (like degree centrality) based on outgoing connections when working with directed networks. In our case, this means centrality scores are initially only calculated for performers (source nodes), since edges represent their participation in films (target nodes). Visual differentiation based on node category is not just a cosmetic enhancement but a critical step in making the network intelligible. While the basic `plot()` function can change the node color based on a value, it is easier to use the "ggraph" package, which handles such intricacies with ease and flexibility.

7.6 ggraph

The "ggraph" package is an extension of the well-established "ggplot2" package we saw in Chap. 4. It inherits the grammar of graphics framework and tailors it to the unique needs of network visualization by introducing new geoms specific to the structural components of a graph, such as nodes, edges, and the paths between them. When we pass a `tbl_graph` object to `ggraph()`, it intelligently recognizes the nature of our data—whether we're working with nodes or edges—and allows us to apply aesthetic mappings accordingly. For example, we can specify that the color of each node geom should correspond to a category attribute within the node tibble or that the width of each edge geom should reflect a weight attribute in the edge tibble.

To begin using "ggraph," it's essential to ensure that the package is installed in our R environment. If it isn't, you can install it using the standard `install.packages("ggraph")` command. Once installed, loading "ggraph" into your R session is as simple as calling `library(ggraph)`.

````
```{r}
# Install the ggraph package
install.packages("ggraph")
# Load the ggraph library for use in current R session
library(ggraph)
```
````

7.7 Network Visualization Layouts

Transitioning from general data visualization to the specialized realm of network graphs entails a pivotal shift in the tools we use, with `ggraph()` replacing the familiar `ggplot()` function. This tailored function is designed specifically for network data, incorporating the complexities and nuances of graph structures into the grammar of graphics framework. A crucial component in crafting a network visualization with `ggraph()` is specifying a layout algorithm, which is set using the `layout` argument within the `ggraph()` function call.

Layout algorithms are sophisticated methods that intelligently position nodes to optimize the readability of the network. They play a critical role in network visualization by resolving the spatial challenges that arise when nodes are plotted. Without a layout algorithm, the nodes in a dense network are likely to overlap, making it difficult to discern individual elements and their connections. The layout algorithm's task is to distribute nodes across the plot in a way that minimizes such overlap and reveals the structure of the network in an uncluttered and interpretable fashion. It's important to note that there is no universally "correct" layout algorithm; rather, each serves as a different lens through which to view the network, and the choice of which to use may vary depending on the researcher's goals.

For our purposes, the "fr" layout, short for Fruchterman-Reingold, is a solid choice for small- to medium-sized datasets. It treats the network like a system of springs and repulsive forces, finding an equilibrium that serves as the foundation for the plot (Fruchterman and Reingold 1991). Still, we will only be plotting the first ten nodes for now. Despite the capabilities of our chosen layout algorithm to arrange nodes in a coherent and visually comprehensible manner, the challenge of visual clutter remains when dealing with larger networks. A multitude of nodes can lead to a dense and convoluted visualization where individual nodes and their connections are obscured, inhibiting our ability to derive meaningful insights from the graph. To start a visualization, you need to call `ggraph()`. However, if this is all you call, you will be greeted with a blank plot. Just like with `ggplot()`, we need to also add layers. To do this, we will add `geom_node_point()` and `geom_edge_link()` to our graph. As the names imply, these will create simple points for each node and a line representing each edge.

```
# Set random seed for consistent node layout in the plot and visualize
↪  the first ten nodes
films_tidygraph |>
  activate(nodes) |>
  slice(1:10) |>
  ggraph(layout = "fr") +
  geom_edge_link() +
  geom_node_point()
```

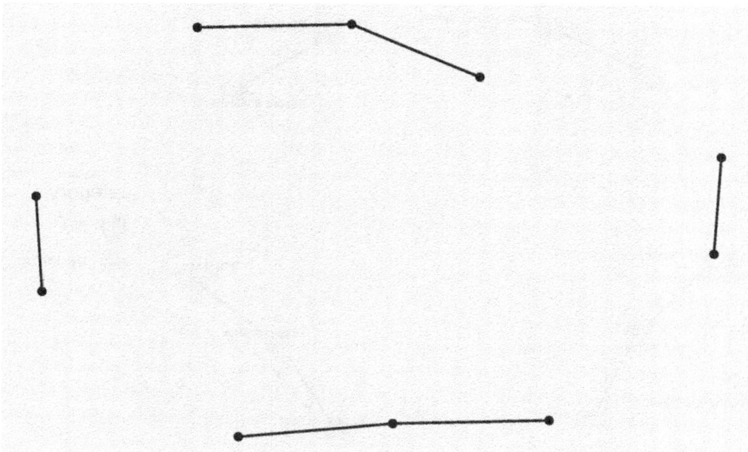

Fig. 7.6 Initial network visualization using ggraph with default node points and edge links

The foundation for our visualization is laid with basic points representing nodes and lines depicting the edges. As seen in Fig. 7.6, however, it is difficult to tell what is going on at this nascent stage. The nodes, in their default rendering, are not only diminutive but lack identifiers that tell us what or whom they represent within the network. To fix this, we need to adjust the aesthetic mappings for our geoms. This is where "ggraph" differs fundamentally from its predecessor, "ggplot2." While "ggplot2" emphasizes a global approach to aesthetic mappings, applying attributes across the entire plot "ggraph" requires a more targeted method due to the dual nature of network data. Node-specific mappings are defined within node geoms, such as `geom-node-point()`, and edge-specific mappings within edge geoms, such as `geom-edge-link()`. This granular approach ensures that the visual attributes of nodes and edges are directly and appropriately mapped, reflecting their distinct characteristics and roles within the network.

Figure 7.7 adds labels, another layer of detail to our nodes. Through `geom-node-text()`, we can overlay textual identifiers on each node, transforming the previously anonymous points into clearly marked entities. However, text labels come with their own challenges, particularly in dense areas of the graph where labels may overlap and become unreadable. The `repel` argument within `geom-node-text()` is designed to address this problem. By setting `repel` to `TRUE`, we activate an algorithm that automatically adjusts the placement of labels to minimize overlap, effectively "repelling" them from one another. This feature ensures that labels remain legible, even in areas where nodes are closely clustered.

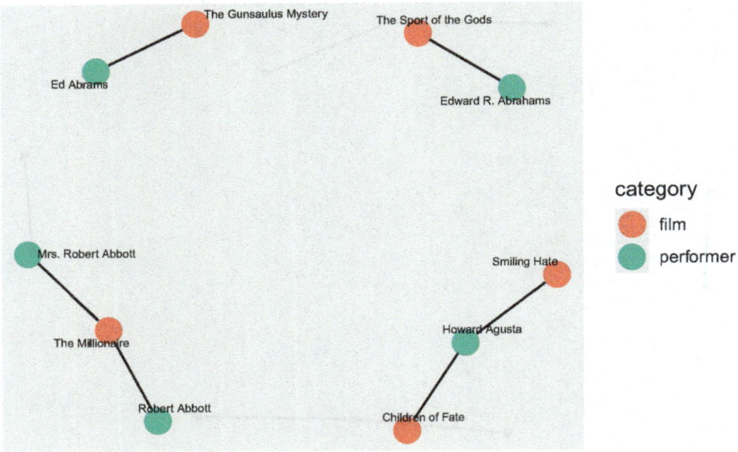

Fig. 7.7 Network visualization with nodes colored by category (performer/film) and labeled

```
# Set random seed for consistent node layout in the plot and visualize
↪ the first ten nodes. Create a label for each node.
films_tidygraph |>
  activate(nodes) |>
  slice(1:10) |>
  ggraph(layout = "fr") +
    geom_edge_link() +
    geom_node_point(aes(color = category), size = 6) +
    geom_node_text(aes(label = name), size = 2, repel = TRUE)
```

7.8 Centrality Measures

The raw data comprising the edges and nodes can provide a wealth of insights on their own, revealing the underlying fabric of the network's architecture through quantitative analysis. However, to distill this information into actionable intelligence—especially when addressing more profound questions of influence and importance within the network—we must delve into the realm of centrality measures.

Centrality measures are a cornerstone of network analysis designed to determine the relative significance of nodes within the network. Each centrality measure illuminates a different aspect of influence and importance, and the appropriate measure to employ will depend largely on the specific research question at hand (Das et al. 2018). Are we interested in identifying the most active participants, the best-positioned intermediaries, or perhaps those with the greatest potential influence? As with visual layouts, the choice of centrality measure must be aligned with the nature of the network and the goals of the analysis.

To begin working with centrality measures, we will focus on degree centrality—one of the most intuitive and commonly used measures—which simply tallies the number of connections a node has. A node with high degree centrality suggests it may be a hub or a central actor within the network. In our study of the race film network, for example, degree centrality allows us to identify the most prolific actors or most performed-in films.

In "tidygraph," we can use the `centrality_degree()` function to calculate this measure for each node. By incorporating this function into a `mutate()` call, we will append a centrality score to each node within our tibble. We can then sort, filter, and visualize this enriched dataset. These insights can inform further research, guide decision-making, or provide historical context about the network's evolution. Let's see how this works in practice:

```
# Add degree centrality scores to nodes in the films_tidygraph object.
films_tidygraph <- films_tidygraph |>
  activate(nodes) |>
  mutate(centrality = centrality_degree())
```

Now that we have calculated the measures for each node, we can use the rest of our dplyr verbs as before to filter and select. Let's arrange by centrality to see the top performers.

```
# Re-activate node view for forthcoming operations and arrange by
↪  centrality
films_tidygraph |>
  activate(nodes) |>
  arrange(desc(centrality))
```

```
# A tbl_graph: 731 nodes and 990 edges
#
# A directed acyclic multigraph with 48 components
#
# Node Data: 731 x 3 (active)
   name               category  centrality
   <chr>              <chr>          <dbl>
 1 Lawrence Chenault performer          22
 2 Samuel Jacks       performer          20
 3 Sam Robinson       performer          20
 4 Will Starks        performer          20
 5 Walter Brogsdale  performer          19
 6 Robert Duree       performer          19
 7 Bert Murphy        performer          19
 8 Mildred Price      performer          19
 9 Evon Skekeeter     performer          19
10 Julia Mason        performer          17
# i 721 more rows
```

```
#
# Edge Data: 990 x 2
   from    to
  <int> <int>
1   118   557
2   119   557
3   120   558
# i 987 more rows
```

Chenault's prominence is no surprise and serves to validate centrality measure as a metric that can capture and quantify the intuitive notions of influence we might have regarding a network's nodes. It also underscores the importance of considering multiple facets of a node's involvement within the network.

In the context of our directed and bimodal network, where nodes represent distinct categories—films and performers—the application of centrality measures introduces a layer of complexity that can lead to some initial confusion. By default, "tidygraph" calculates centrality measures such as degree centrality based on outgoing connections, which, in our case, corresponds to the performers, since the edges define their participation in films. This focus on outgoing connections stems from the directed nature of the network, where edges have a specific direction that indicates the relationship dynamic from one node type to another.

This default behavior results in centrality measures being displayed only for performers, as they are the source nodes in the "performed in" relationship. Films, on the other hand, do not show centrality measures, because they are the target nodes—the endpoint of the directed edges. However, the significance of films within the network is equally important to understand. Their centrality could reflect their notability within the industry, such as the number of prominent actors associated with a particular film or a film's role in bringing together diverse talent.

To address this and obtain a comprehensive view of the network's dynamics, we can adjust the calculation of centrality measures to include target nodes by changing the `mode` argument. Setting `mode` to "in" recalculates the centrality measures based on incoming connections, which would highlight films that have the most performers associated with them. This adjustment allows us to consider the centrality of films within the network, revealing those that serve as crucial hubs of collaboration.

For a holistic analysis that takes into account the centrality of both performers and films, we can set the `mode` argument to "all". This approach combines both incoming and outgoing connections, providing a centrality score that captures the overall connectivity of a node regardless of type. By doing so, we can discern which nodes—be they performers or films—are most central in terms of their influence on others and their reception of influence.

```
# Calculate degree centrality for all nodes in the films_tidygraph
↪  object, considering both incoming and outgoing connections
films_tidygraph <- films_tidygraph |>
  activate(nodes) |>
  mutate(centrality = centrality_degree(mode = "all")) # Consider all
  ↪  connections
```

Let's now look at our top 15 nodes.

```
# Re-activate node view for forthcoming operations and arrange by
↪  centrality
films_tidygraph |>
  activate(nodes) |>
  arrange(desc(centrality))
```

```
# A tbl_graph: 731 nodes and 990 edges
#
# A directed acyclic multigraph with 48 components
#
# Node Data: 731 x 3 (active)
   name              category  centrality
   <chr>             <chr>          <dbl>
 1 Deceit            film              25
 2 Lawrence Chenault performer         22
 3 Samuel Jacks      performer         20
 4 Sam Robinson      performer         20
 5 Will Starks       performer         20
 6 Walter Brogsdale  performer         19
 7 Robert Duree      performer         19
 8 Bert Murphy       performer         19
 9 Mildred Price     performer         19
10 Evon Skekeeter    performer         19
# i 721 more rows
#
# Edge Data: 990 x 2
   from    to
  <int> <int>
1   258    35
2   259    35
3   260    20
# i 987 more rows
```

Having computed centrality measures for both films and performers in our directed and bimodal network, we are equipped with a valuable set of data points that reflect the connectivity and potential influence of each node within the network. However, it's important to understand that combining these metrics across nodes of different types can lead to misinterpretation and an unclear representation of the network's true dynamics.

The crux of the issue lies in the fundamentally different roles that films and performers play within the network. For performers, centrality may reflect their prolific nature or their prominence within the industry, whereas for films, centrality could indicate the extent to which a film brings together various performers or serves as a focal point within the network. These differing implications suggest that centrality measures are context-dependent and should be considered within the scope of what each node type represents.

To underscore why it is not advisable to directly combine these disparate centrality metrics, visualizing the results can be illuminating. Let's manipulate the size of the nodes based on our centrality degree. We will then look at the top 20 nodes by degree.

```
## Create a network visualization of the top 20 nodes by degree
↪ centrality
films_tidygraph |>
  activate(nodes) |>
  slice(1:20) |>
  ggraph(layout = "fr") +
    geom_edge_link() +
    geom_node_point(aes(color = category, size = centrality)) +
    geom_node_text(aes(label = name), repel = TRUE, size = 2)
```

Figure 7.8 visualizes the top 20 nodes by degree centrality. For someone looking at the network visualization without a deep understanding of its bimodal structure, it might be tempting to interpret the size of the nodes as a uniform measure of

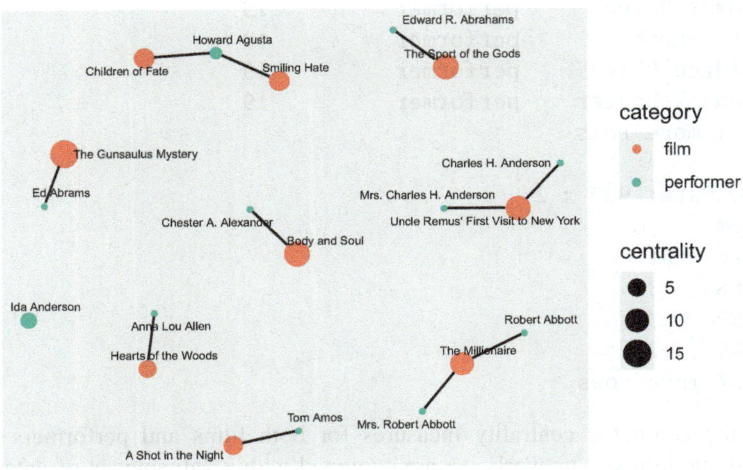

Fig. 7.8 Network visualization of the top 20 nodes ranked by degree centrality, sized accordingly

influence or importance across the entire network. This could lead to the erroneous assumption that a film with a large node, such as *Deceit*, is necessarily more influential than an individual performer with a smaller node, like Ida Anderson. However, as we know, the node sizes are indicative of two different kinds of centrality within the distinct categories. A larger node for a film does not necessarily mean it was more influential in the broader context of the industry but rather that it was a significant point of convergence for actors within the network. Conversely, the node size for performers measures the breadth of their work, which could be an indicator of their demand, versatility, or popularity in the race film sphere.

Using our color-coding scheme where, for instance, all performer nodes are shaded one color, while film nodes are shaded another helps. However, color, shape, and labels, while clarifying, don't address the core issue of comparing centrality across two fundamentally different types of entities. This is where the concept of bipartite projection becomes invaluable.

7.9 Bipartite Graphs

Bipartite projection is a powerful technique that restructures a bimodal graph into two unimodal graphs, each representing connections within a single category of nodes (Zhou et al. 2007). For our analysis of race films, this means translating the complex web of relationships between performers and films into two separate, more focused networks that can reveal the underlying patterns of collaboration.

Let's think about how this works with the data we have for performers. The first step in this process is to establish whether a pair of performers appeared together in the same film. Next, we create a new network that encapsulates these connections, effectively building a map of performer-to-performer relationships. In this new graph, nodes represent individual actors, and edges represent their coappearances in films. The third step involves quantifying these relationships by calculating the number of connections between each performer. This degree count serves as an indicator of how intertwined each actor's career is with their peers, providing insight into the collaborative landscape of the industry. Finally, we visualize this unimodal network, with the node size adjusted to reflect the degree of connectivity for each performer with larger nodes indicate actors with more extensive collaborative histories.

The "igraph" package in R simplifies this process with its `bipartite.projection()` function. Since "tidygraph" is built on top of "igraph," our network object, it is compatible with "igraph's" suite of tools. By employing the `class()` function, we can confirm this compatibility, as it reveals that our "tidygraph" object is indeed an "igraph" object under the hood. This interoperability is particularly useful, allowing us to leverage the specialized capabilities of "igraph" directly on our "tidygraph" object without the need for complex conversions or additional data manipulation.

```
# Check Class for Tidygraph Object
class(films_tidygraph)
```

```
[1] "tbl_graph" "igraph"
```

The process of bipartite projection using "igraph" can be a little counterintuitive, because it introduces a particular requirement: The dataset must include a column explicitly named `type`, which functions as a distinguishing marker to differentiate the two distinct sets of nodes that constitute the bipartite network. Each node must be tagged either TRUE or FALSE. In the context of our analysis, where the network consists of performers and films, we designate performers with a FALSE label and films with a TRUE label. However, this categorization is arbitrary in terms of the algorithm's function; you could feasibly reverse these labels (i.e., performers as TRUE and films as FALSE) and still arrive at the same bipartite projection results.

Assigning these binary labels to our dataset is achieved through a combination of "dplyr's" `mutate()` function and the `case_when()` function. The 'igraph' package requires a logical `type` column where one node type is marked TRUE and the other FALSE. In our case, we mark films as TRUE and performers as FALSE. This distinction is arbitrary for the algorithm's purposes (we could reverse them) but helps maintain consistency in our analysis. By utilizing `case_when()` within `mutate()`, we can create logical conditions that automatically assign TRUE to films and FALSE to performers based on these criteria.

```
# Create 'type' column that labels node types with TRUE/FALSE for
↪  bipartite functionality
films_tidygraph <- films_tidygraph |>
  activate(nodes) |>
  mutate(type = case_when(
    category == "film" ~ TRUE,
    category == "performer" ~ FALSE
  ))

# View updated tidygraph
films_tidygraph
```

```
# A tbl_graph: 731 nodes and 990 edges
#
# A directed acyclic multigraph with 48 components
#
# Node Data: 731 x 4 (active)
  name                category  centrality type
  <chr>               <chr>          <dbl> <lgl>
1 Robert Abbott       performer          1 FALSE
2 The Millionaire     film              10 TRUE
3 Mrs. Robert Abbott  performer          1 FALSE
4 Edward R. Abrahams  performer          1 FALSE
```

```
 5 The Sport of the Gods film                   12 TRUE
 6 Ed Abrams              performer              1 FALSE
 7 The Gunsaulus Mystery film                   15 TRUE
 8 Howard Agusta          performer              2 FALSE
 9 Smiling Hate           film                   7 TRUE
10 Children of Fate       film                   7 TRUE
# i 721 more rows
#
# Edge Data: 990 x 2
   from    to
  <int> <int>
1     1     2
2     3     2
3     4     5
# i 987 more rows
```

To make sure that, we do indeed have a bipartite graph. We can use the `is_bipartite()` function. However, before we do, we should make sure we have installed and loaded the "igraph" package.

```
```{r}
# Installs the igraph package from CRAN
install.packages("igraph")
# Loads igraph library for use in current R session
library(igraph)
```
```

Now that we have installed and loaded "igraph," we can use the `is_bipartite()` function:

```
# Assure that the graph is bipartite
is.bipartite(films_tidygraph)
```

```
[1] TRUE
```

As the output shows, we do indeed have a bipartite graph. To cast it into two separate unimodal graphs, we simply pass our graph to "igraph's" `bipartite.projection()` function. Remember that this will give us two different "projections." One projection will be for actors, and the other one will be for the films themselves.

```
# Perform bipartite projection on the tidygraph object
films_projection <- bipartite.projection(films_tidygraph)

# Output projections: performer network (proj1) and film network (proj2)
films_projection
```

```
$proj1
IGRAPH 6c4e4a8 UNW- 556 2945 --
+ attr: name (v/c), category (v/c), centrality (v/n), weight (e/n)
+ edges from 6c4e4a8 (vertex names):
 [1] Robert Abbott     --Mrs. Robert Abbott
 [2] Robert Abbott     --Vera Bracken
 [3] Robert Abbott     --J. Lawrence Criner
 [4] Robert Abbott     --Cleo Desmond
 [5] Robert Abbott     --William Edmondson
 [6] Robert Abbott     --Samuel Jacks
 [7] Robert Abbott     --Lionel Monagas
 [8] Robert Abbott     --Grace Smith
+ ... omitted several edges

$proj2
IGRAPH 310f6d7 UNW- 175 844 --
+ attr: name (v/c), category (v/c), centrality (v/n), weight (e/n)
+ edges from 310f6d7 (vertex names):
 [1] The Millionaire--The Brute
 [2] The Millionaire--The Flying Ace
 [3] The Millionaire--Black Gold
 [4] The Millionaire--Oft in the Silly Night
 [5] The Millionaire--Deceit
 [6] The Millionaire--The Midnight Ace
 [7] The Millionaire--Thirty Years Later
 [8] The Millionaire--Are Working Girls Safe?
+ ... omitted several edges
```

When examining the output generated by the "igraph" package after the bipartite projection process, we encounter a structural representation that deviates from conventional data frames or tibbles we have been handling. As we delve into the results of the bipartite projection, it becomes evident that "igraph" has encapsulated the outcome within a list format. The first list element, aptly named "proj1," is a graph object that represents only the performers. Each node within this graph signifies an individual performer, and the edges denote the shared experiences of appearing in the same films. Similarly, the second list element, labeled "proj2," is solely dedicated to the films. Here, each node corresponds to a specific film, and the edges reflect the presence of shared performers, illustrating a network where films are interlinked based on common cast members.

To make it easier to work with our data, we can now convert these back into two separate "tidygraph" objects. This is easy to do using the `as_tbl_graph()` function.

```
# Convert unimodal graph for films to tidygraph and view
performer_projection <- as_tbl_graph(films_projection$proj1)
    # Convert unimodal graph for actors to tidygraph and view
    film_projection_graph <- as_tbl_graph(films_projection$proj2)
```

Let's now look at the output of our performer_projection.

```
# View the performer projection graph
performer_projection
```

```
# A tbl_graph: 556 nodes and 2945 edges
#
# An undirected simple graph with 48 components
#
# Node Data: 556 x 3 (active)
   name                      category  centrality
   <chr>                     <chr>          <dbl>
 1 Robert Abbott             performer          1
 2 Mrs. Robert Abbott        performer          1
 3 Edward R. Abrahams        performer          1
 4 Ed Abrams                 performer          1
 5 Howard Agusta             performer          2
 6 Chester A. Alexander      performer          1
 7 Anna Lou Allen            performer          1
 8 Tom Amos                  performer          1
 9 Charles H. Anderson       performer          1
10 Mrs. Charles H. Anderson performer          1
# i 546 more rows
#
# Edge Data: 2,945 x 3
   from    to weight
  <int> <int>  <dbl>
1     1     2      1
2     1    39      1
3     1   109      1
# i 2,942 more rows
```

Having completed our projection, we can now arrange by the top edges by weight to unravel the layers of relational data and can identify the most frequent collaborations or the strongest partnerships within the performer network.

```
# Sort edges by weight
performer_projection |>
  activate(edges) |>
  arrange(desc(weight))
```

```
# A tbl_graph: 556 nodes and 2945 edges
#
# An undirected simple graph with 48 components
#
```

```
# Edge Data: 2,945 x 3 (active)
    from    to weight
   <int> <int>  <dbl>
 1    41   147     19
 2    41   390     19
 3    41   408     19
 4    41   434     19
 5    41   456     19
 6   147   390     19
 7   147   408     19
 8   147   434     19
 9   147   456     19
10   243   456     19
# i 2,935 more rows
#
# Node Data: 556 x 3
  name                category  centrality
  <chr>               <chr>          <dbl>
1 Robert Abbott       performer          1
2 Mrs. Robert Abbott  performer          1
3 Edward R. Abrahams  performer          1
# i 553 more rows
```

At present, our data presents us with a quandary: The identity of the nodes linked to the highest weight scores remains obscured, as they are currently disconnected from their defining nodal attributes. Without merging the weight information from our edges with the individual characteristics of each node, we're left with a picture that lacks clarity and specificity. We can use "ggraph" to create a visual representation of our network that highlights the edges with the highest weight scores and annotates them with the corresponding performer names. Let's take a look at the top 20 collaboration.

```
# Sort edges by weight and visualize the top 20 edges
performer_projection |>
  activate(edges) |>
  arrange(desc(weight)) |>
  slice(1:20) |>
  ggraph(layout = "fr") +
    geom_edge_link(aes(color = weight)) +
    geom_node_point() +
    geom_node_text(aes(label = name), repel = TRUE, size = 2)
```

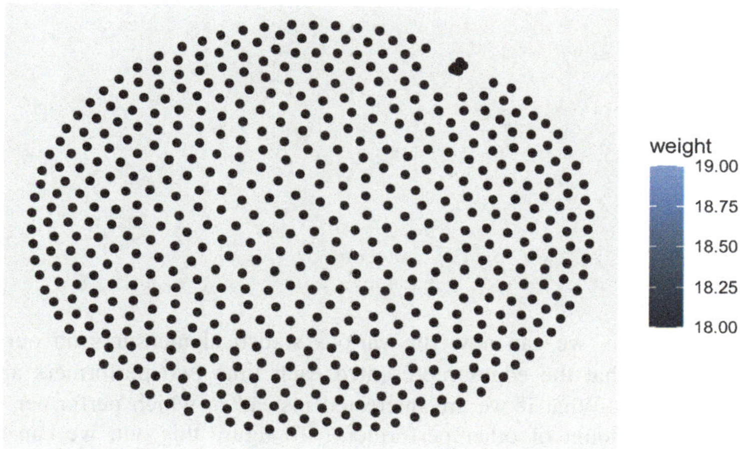

Fig. 7.9 Network visualization highlighting the top 20 performer collaborations (edges colored by weight), including potentially isolated nodes

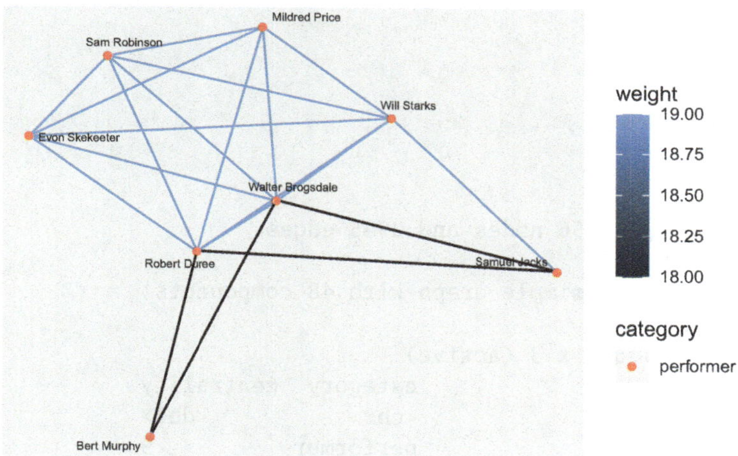

Fig. 7.10 Network visualization of the top 20 performer collaborations after removing isolated nodes

Figure 7.9 illustrates that it is very hard to tell what is going on here because of the large amount of nodes. But why did these nodes show up when we filtered out our edges? In a network graph, it is perfectly valid to have nodes without edges. Consequently, when we filter our edges, "tidygraph" does not remove the isolated nodes. To remove nodes that are isolated, as shown in Fig. 7.10, we can filter them and use the `node-is-isolated()` function.

```
# Filter out isolated nodes and visualize the top 20 edges
performer_projection |>
  activate(edges) |>
  arrange(desc(weight)) |>
  slice(1:20) |>
  activate(nodes) |>
  filter(!node_is_isolated()) |>
  ggraph(layout = "fr") +
    geom_edge_link(aes(color = weight)) +
    geom_node_point(aes(color = category)) +
    geom_node_text(aes(label = name), repel = TRUE, size = 2)
```

Just like before, we can now run various statistical measures on our graph. Above, we saw that the edges represented each time two performers appeared in a film together. What if we are interested instead in which performer worked with the most amount of other performers? To figure this out, we can use the `centrality_degree()` function with `mutate()` in the same way we saw earlier.

```
# Perform centrality measure within the projected exclusive performer
↪ nodes
performer_projection <- performer_projection |>
  activate(nodes) |>
  mutate(centrality = centrality_degree())

# Display the updated graph object, including new 'centrality' values
performer_projection
```

```
# A tbl_graph: 556 nodes and 2945 edges
#
# An undirected simple graph with 48 components
#
# Node Data: 556 x 3 (active)
   name                     category  centrality
   <chr>                    <chr>            <dbl>
 1 Robert Abbott            performer            9
 2 Mrs. Robert Abbott       performer            9
 3 Edward R. Abrahams       performer           11
 4 Ed Abrams                performer           14
 5 Howard Agusta            performer           10
 6 Chester A. Alexander     performer           11
 7 Anna Lou Allen           performer            4
 8 Tom Amos                 performer            5
 9 Charles H. Anderson      performer           10
10 Mrs. Charles H. Anderson performer           10
# i 546 more rows
#
# Edge Data: 2,945 x 3
```

```
   from    to weight
  <int> <int>  <dbl>
1     1     2      1
2     1    39      1
3     1   109      1
# i 2,942 more rows
```

We can now visualize the results using "ggraph." To avoid having too many nodes, we will filter by those performers that are connected to at least forty other performers. In addition, because differentiation of nodes can be difficult when the size is only used, we will also use color to make the graph easier to see.

```
# Visualize the performer projection graph with centrality scores
performer_projection |>
  activate(nodes) |>
  filter(centrality>40) |>
  ggraph(layout = "fr") +
    geom_edge_link() +
    geom_node_point(aes(size = centrality, color = centrality)) +
    geom_node_text(aes(label = name), repel = TRUE, size = 2)
```

Figure 7.11 shows that the performer who has worked with the most other performers is Lawrence Cheault, followed by Evelyn Preer and then E.G. Tatum.

Remember that the total number of degrees is only one of many centrality measures and is not always representative of the most important node in the network, depending on your definition. For instance, while you may be connected to many other individuals, your ability to traverse and pass information through the network may be limited. Another measure we could do is calculate eigenvector centrality,

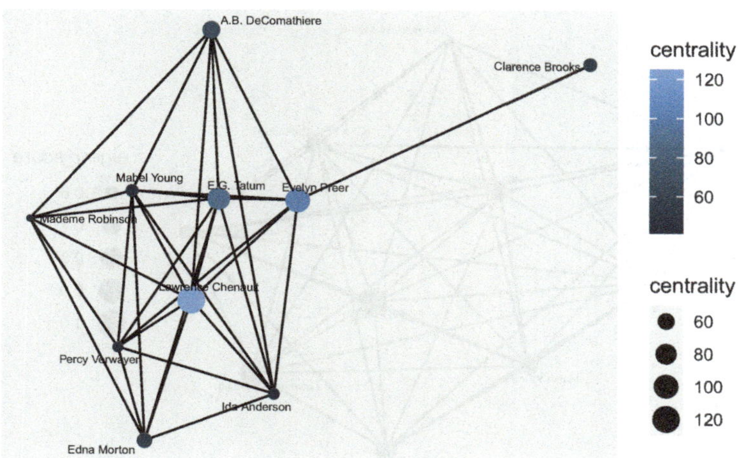

Fig. 7.11 Network visualization showing performers with degree centrality greater than 40

which measures how connected a node is to other nodes that are well connected. The only change we need to make is to use `centrality_eigen()` in `mutate()`.

```
# Initialize performer_projection for upcoming centrality calculation.
↪  Calculate eigenvector centrality
performer_projection <- performer_projection |>
  activate(nodes) |>
  mutate(eigen_score = centrality_eigen())
```

Just as before, we can now visualize the results:

```
# Visualize the performer projection graph with eigenvector centrality
↪  scores
performer_projection |>
  activate(nodes) |>
  filter(eigen_score>0.5) |>
  ggraph(layout = "fr") +
    geom_edge_link() +
    geom_node_point(aes(size = eigen_score)) +
    geom_node_text(aes(label = name), repel = TRUE, size = 2)
```

As seen in Fig. 7.12, E.G. Tatum's preeminent eigenvector score suggests that he is not just well-connected across the network but is also linked to other highly connected and influential nodes. This could indicate that he frequently worked with other prominent actors or that he appeared in a number of significant films that are central within the network of race films. It might also suggest that his presence in a film could elevate the film's status within the network, attracting other influential actors to the project, or that his collaborations with other actors were significant enough to enhance their own standing within the industry.

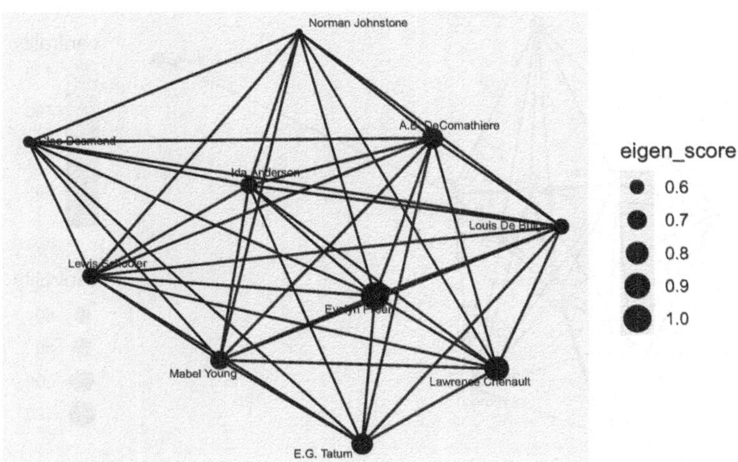

Fig. 7.12 Network visualization showing performers with eigenvector centrality greater than 0.5

7.10 Community Detection

Along with detecting centrality, we can also use "tidygraph" to look at communities that emerge in the network. The concept of a community within network theory is multifaceted and nuanced. One might initially assume that a community is simply a collection of the most densely interconnected nodes, where each member is directly tied to many others within the group. Alternatively, one might consider a community to be a cluster of nodes that, while possibly less connected internally, are distinctly isolated or segregated from the rest of the network. However, the reality is that both these definitions can hold true depending on the context and the specific characteristics of the network under examination.

Community detection, therefore, is the process of identifying these groups within a larger network, and several algorithms exist to address this complex task. Each algorithm brings a different perspective to what constitutes a community and employs unique methodologies to uncover these structures.

We will use the Louvain method of community detection as it is widely recognized for its efficiency and effectiveness in large-scale networks. The Louvain method operates on the principle of modularity optimization. It iteratively groups nodes into communities in a way that maximizes the density of connections within communities compared to connections between communities. In essence, it seeks to draw borders around clusters of nodes that interact more frequently with each other than with those outside their group. This approach can be advantageous due to its scalability and relative simplicity. However, it's important to recognize that like all algorithms, it is not without its limitations. For instance, it may not always detect smaller communities within a network or may be influenced by the resolution limit, potentially affecting the granularity of the resulting community divisions.

The process for doing community detection in "tidygraph" is similar to the process of calculating node metrics. We will use the `group_louvain()` function with `mutate()` to create a new column for each node representing its community (Blondel et al. 2008). As before, we are going to set the seed to assure reproducibility. If you don't do this step and did not do it previously, your output may differ.

```
# Assign nodes to communities using the Louvain algorithm.
performer_projection <- performer_projection |>
  activate(nodes) |>
  mutate(community = group_louvain())
```

Let's look at the output.

```
# Check projection output structure and content
performer_projection
```

```
# A tbl_graph: 556 nodes and 2945 edges
#
# An undirected simple graph with 48 components
#
```

```
# Node Data: 556 x 5 (active)
   name                       category  centrality eigen_score community
   <chr>                      <chr>          <dbl>       <dbl>     <int>
 1 Robert Abbott              performer          9      0.0959         3
 2 Mrs. Robert Abbott         performer          9      0.0959         3
 3 Edward R. Abrahams         performer         11      0.0740         2
 4 Ed Abrams                  performer         14      0.219          5
 5 Howard Agusta              performer         10      0.0578         1
 6 Chester A. Alexander       performer         11      0.119          1
 7 Anna Lou Allen             performer          4      0                 24
 8 Tom Amos                   performer          5      0                  6
 9 Charles H. Anderson        performer         10      0.0165        13
10 Mrs. Charles H. Anderson performer         10      0.0165        13
# i 546 more rows
#
# Edge Data: 2,945 x 3
   from    to weight
  <int> <int>  <dbl>
1     1     2      1
2     1    39      1
3     1   109      1
# i 2,942 more rows
```

Upon completion of the community detection process using the Louvain method, our dataset has evolved to include a pivotal new column. Each node within this column is now associated with a distinct community, effectively grouping individuals or entities based on the network's dense interconnections. This emergent column offers a fresh lens through which to analyze our network and paves the way for targeted analysis. For instance, one might investigate the centrality measures within each community to identify key influencers or pivotal nodes that hold the community together. Additionally, examining the nodes that bridge communities can shed light on the flow of information or resources and highlight potential areas for intervention or enhancement.

7.11 Conclusion

In this chapter, we ventured into the realm of network analysis, wielding the powerful toolkit provided by "tidygraph" to navigate the complex interconnections within cultural data. By extending the familiar techniques and functions of the "tidyverse" into network theory, "tidygraph" has enabled us to approach our data with both finesse and depth. We have seen how centrality measures illuminate the most influential nodes within a network and how they can also reveal intricate clusters that might represent subcultures or tightly knit collaborative groups.

The exploratory power of the grammar of graphics, harnessed through "ggraph," has opened up new avenues for visual storytelling, transforming the abstract into the concrete. Through this lens, we created visualizations that not only depicted network structures but also conveyed the weight of relationships and the significance of individual nodes. This dual capacity for analysis and communication is indispensable for cultural analysts who seek to interpret and share complex data narratives.

As we continue to delve into vast troves of data, network theory challenges us to reimagine what can be visualized as a network. Yet, it is crucial to remain cognizant of both the strengths and limitations inherent in network visualizations. While networks can bring clarity to complexity, they may also oversimplify or fail to capture the dynamism of temporal changes. As cultural analysts, it behooves us to continually evaluate the tools at our disposal, to question the appropriateness of our models, and to consider the implications of our analytical choices. With these considerations in mind, network theory stands out as a potent instrument in our analytical repertoire, one that can uncover patterns and stories within cultural data that might otherwise remain untold.

Chapter 8
Machine Learning with Tidy Models

Abstract This chapter explores machine learning within the tidyverse in R. It details "tidymodels," a "meta-package" that streamlines various machine learning workflows, facilitating the use of advanced analytical methods in cultural research. A key focus is placed on decision trees, random forests, and the critical role of hyperparameter tuning. Using a case study from art history, the chapter demonstrates how these methods can uncover biases in the representation of artists across demographics in educational textbooks. Through guided examples, readers gain insights into the iterative nature of model tuning using hyperparameter grids and learn how to assess the accuracy of predictive models.

Keywords Machine learning · Tidyverse · `tidymodels` · Cultural analytics · Decision trees · Random forests · Hyperparameter tuning · Predictive modeling · Model evaluation

8.1 The Rise of Tidy Models

Some of the most interesting developments in cultural analytics center around topics related to "machine learning." Nonetheless, the definition of what constitutes machine learning is ambiguous at best, often leading to misunderstanding among researchers and practitioners. We have already seen some examples of machine learning in previous chapters when we used linear regression to identify patterns and predict outcomes. While linear regression is technically machine learning, the term is more commonly associated with other complex analytical methods and algorithms, such as decision trees, random forests, support vector machines, and deep neural networks. Each of these has its strengths and weaknesses, and knowing which one to use in a given task is crucial for obtaining accurate and meaningful results.

Because of the myriad of machine learning techniques available, the R community has had a difficult time developing an underlying philosophy or framework for how to implement these methods in a consistent and streamlined way. One of the earliest packages to try to solve this issue was the `caret` package created by

N. Siddiqui, *Cultural Analytics in R: A Tidy Approach*, Use R!,
https://doi.org/10.1007/978-3-031-96618-7_8

Max Kuhn (2008). This package benefited from providing a uniform framework for individuals to complete different steps of the machine learning pipeline such as preprocessing data, performing cross-validation, and adjusting hyperparameters. In addition, it made it easy to visualize and evaluate model performance which was incredibly valuable for fine-tuning model settings and improving accuracy.

Although the "caret" package was revolutionary for R, it also had limitations in terms of syntax and performance. Since it applied its own implementation of various algorithms, the results were not always compatible with other packages. Keeping up with the latest advancements became challenging, and the rise of the "tidy" paradigm meant many users sought a package that had a stronger focus on integration with the wider tidyverse ecosystem. This resulted in a "meta-package" entitled "tidymodels," developed by Julia Silge and Max Kuhn (former lead developer of the "caret" package), which sought to continue the legacy of "caret" while addressing its constraints (Kuhn and Wickham 2020).

`tidymodels`, like `caret`, seeks to offer a cohesive collection of packages aimed at streamlining machine learning workflows. It provides a grammar for modeling that is both consistent with the ideology of the tidyverse and flexible enough to model a wide array of data science tasks. By abstracting away the complexities of different modeling functions and algorithms, the package allows researchers to focus more on the interpretation and storytelling aspects of their data, which are central to cultural analytics. This abstraction means switching between different modeling approaches does not require learning new syntax or functions, making exploratory analysis more fluid and intuitive. Additionally, the modularity of `tidymodels` allows for easy expansion and customization to meet the specific needs of cultural analytics, because new modeling techniques or data preprocessing steps can be integrated into the existing workflow without disrupting the existing structure.

In this chapter, we will focus only on a few foundational methods that are widely used and available within the "tidymodels" framework but which have, as of yet, played little role in cultural analytics due to their perceived complexity or computational demands: decision trees. Because decision trees are relatively straightforward, they serve as excellent introductory examples for those new to machine learning. We will focus on building a solid understanding of the core features and functions of "tidymodels," such as data preprocessing, hyperparameter tuning, and model evaluation. We will also learn how "tidymodels" simplifies complex algorithms and their applications by observing how minor adjustments to our code can allow us to investigate other machine-learning techniques using the package's versatility.

To see how knowing these techniques help us in practical terms, we will explore a comprehensive dataset meticulously compiled by Holland Stam for her undergraduate thesis (Stam n.d.). Stam's research provides a window into the representation of artists within two seminal art history textbooks: H. W. Janson's *History of Art* and Helen Gardner's *Art Through the Ages*. These texts have achieved widespread acclaim and usage across educational institutions globally, often serving as the cornerstone of art history curricula. By examining these influential books, Stam aimed to illuminate the relationship between demographic attributes of artists—such as gender, race, and nationality—and the extent of their representation in these educational materials. While the coverage of minority artists has increased throughout the 8 editions of Janson's textbook and 16 editions of Gardner's textbook, there is still a significant overrepresentation of white male non-Hispanic or Latinx artists.

We will expand on Stam's research to better understand the patterns of representation and possibly uncover biases that persist in the material. We will see if decision trees and random forests can help us better analyze whether there is a correlation between an artist's demographic profile and the space allocated to their work in the textbooks. Furthermore, we will scrutinize the intersection of textbook representation with the broader art world, particularly by examining the frequency of exhibitions for these artists at prominent institutions like the Museum of Modern Art (MoMA) and the Whitney Museum of American Art. The goal is not only to inform future editions of art history textbooks but also to provoke a critical conversation about diversity and representation in the art historical narrative.

8.2 Loading the Art History Dataset

As we have done in previous chapters, we will load in our dataset using the `read_csv()` function and look at the results using `glimpse()`. Furthermore, we will remove any columns that have NAs, so that our data remains as clean and robust as possible. This preprocessing step is crucial, because machine learning algorithms require complete cases to operate effectively.

```
# Read the artists dataset from a CSV file
artists <- read_csv("data/ch7_artists.csv")

# Remove rows with missing values in any column to ensure complete cases
↪ for subsequent examples
artists <- artists |>
  na.omit()

# Display a summary of the artists dataset
glimpse(artists)
```

```
Rows: 3,104
Columns: 14
$ artist_name                  <chr> "Aaron Douglas", "Aaron Douglas", "Aaron Do~
$ edition_number               <dbl> 9, 10, 11, 12, 13, 14, 15, 16, 14, 15, 16, ~
$ year                         <dbl> 1991, 1996, 2001, 2005, 2009, 2013, 2016, 2~
$ artist_nationality           <chr> "American", "American", "American", "Americ~
$ artist_nationality_other     <chr> "American", "American", "American", "Americ~
$ artist_gender                <chr> "Male", "Male", "Male", "Male", "Male", "Ma~
$ artist_race                  <chr> "Black or African American", "Black or Afri~
$ artist_ethnicity             <chr> "Not Hispanic or Latino origin", "Not Hispa~
```

```
$ book                         <chr> "Gardner", "Gardner", "Gardner", "Gardner",~
$ space_ratio_per_page_total <dbl> 0.3533658, 0.3739470, 0.3032593, 0.3770489,~
$ artist_unique_id             <dbl> 2, 2, 2, 2, 2, 2, 2, 2, 4, 4, 4, 6, 6, 6, 6~
$ moma_count_to_year           <dbl> 0, 0, 0, 0, 0, 0, 0, 0, 0, 0, 0, 0, 0, 0, 0~
$ whitney_count_to_year        <dbl> 0, 0, 0, 0, 0, 0, 0, 0, 0, 0, 0, 0, 0, 0, 0~
$ artist_race_nwi              <chr> "Non-White", "Non-White", "Non-White", "Non~
```

As you can see, we have a little over 3100 different data points, and for each data point, we have the following variables:

artist_name: The name of the artist involved in the project or artwork.
edition_number: The edition number of the book the artist coverage was measured in.
year: The publication year of the book being examined.
artist_nationality: The nationality of the artist covered.
artist_nationality_other: According to Stam, roughly 80 percent of the individuals surveyed were either French, Spanish, British, American, or German. Therefore, the categorical strings of this variable are French, Spanish, British, American, German, and others.
artist_gender: The binary gender of the artist examined.
artist_race: The race of the artist examined.
artist_ethnicity: The ethnicity of the artist examined.
space_ratio_per_page_total: The area representing the total amount of space that both the text and the figure of a particular artist take up, as a ratio to the total area per page in the book.
book: Whether the artist data came from Janson's *History of Art* or Gardner's *Art Through the Ages.*
moma_count_to_year: Represents the number of times that the artist has had their work displayed in the Museum of Modern Art in New York City in the given year.
whitney_count_to_year: Represents the amount of times that the artist has had their work displayed in the Whitney Museum in New York City in the given year.
artist_race_nwi: Given the lack of racial diversity, Stam constructed a variable referred to as NWI (Non-White Identity), which rearranges artists into two categories–those who do not affiliate with white racial identity, and those who do.

8.3 Exploratory Analysis of the Art History Dataset

As always, we will do some preliminary data analysis to get a good foundation for our data. As a starting point, we will visualize the distribution of artists' genders within the dataset through a bar graph. This initial step will reveal any gender imbalances that could influence our subsequent findings and provide a clear visual

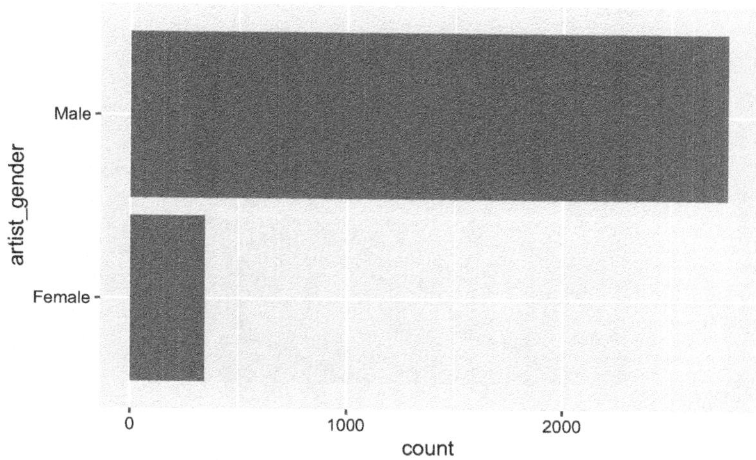

Fig. 8.1 Distribution of artist genders in the art history textbook dataset

representation of the current state of gender diversity in art history textbooks. We will create a bar graph that quantifies and compares the number of male and female artists represented in Janson's *History of Art* and Gardner's *Art Through the Ages*.

```
# Create a horizontal bar plot of artist gender
artists |>
  ggplot(aes(x=artist_gender)) +
  geom_bar() +
  coord_flip()
```

Figure 8.1 shows that there is a pronounced skew toward male artists within the dataset. This majority representation of male artists underscores a significant gender imbalance that warrants further scrutiny. While at first glance this pattern appears to hold true across both textbooks, we may want to probe deeper to ascertain whether the distribution is consistent or if there are underlying differences between the two texts.

```
# Create a bar plot with position set to "fill" for proportional
↪  representation
artists |>
  ggplot(aes(x=book, fill=artist_gender)) +
  geom_bar(position = "fill") +
  coord_flip()
```

Figure 8.2 paints a stark picture—both Gardner and Janson have historically skewed toward featuring more male than female artists in their respective textbooks. This imbalance prompts further investigation, particularly in light of the recent push within the field of art history to broaden the diversity of artists covered, taking into

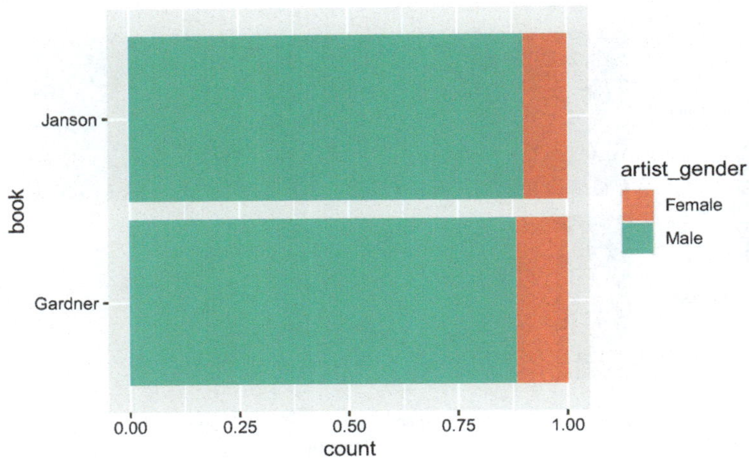

Fig. 8.2 Proportional representation of artist genders within each textbook (Gardner vs. Janson)

account not just gender but also race and nationality. Such an initiative is reflective of a growing awareness and commitment to inclusivity, recognizing the rich and varied contributions of artists from different backgrounds to the tapestry of art history. Let's see how well this has been reflected in the scope of art featured in these textbooks over time. We can do this by simply adding the year to the x-axis and faceting by the book. Due to the fact that Janson had a 1.1 edition, the chart layout can be difficult to read if both the charts follow the same scale. So, we will set the `facet-wrap()` scale parameter to "free." In addition, we will filter out the 1.1 edition, which has little difference in terms of the parameters that we are interested in. However, feel free to include it if you want to explore the data more in depth.

```
# Exclude edition 1.1 for clarity
artists |>
  filter(edition_number != 1.1) |>
  ggplot(aes(x = edition_number, fill = artist_gender)) +
  geom_bar(position = "fill") +
  facet_wrap(~book, scale = "free") +
  coord_flip()
```

Figure 8.3 shows that while it's heartening to observe an upward trend in the visibility of female artists over the years, their proportion still pales in comparison to their male counterparts. The graphs distinctly show that despite a period of positive change, recent editions of both texts have reached a plateau in gender diversity, with male artists maintaining their historical predominance.

The inquiry into gender representation leads us to another critical dimension of diversity: the representation of race within these art history textbooks. Just as we examined gender, it is equally important to analyze how artists of different racial backgrounds are portrayed—or potentially omitted—in these educational resources.

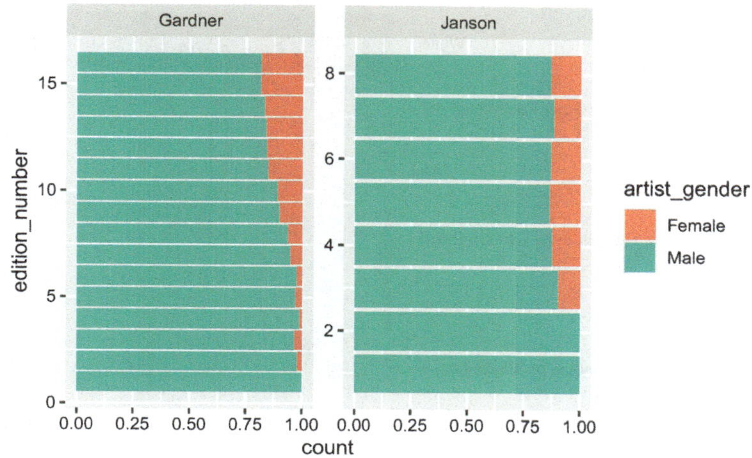

Fig. 8.3 Proportional representation of artist genders across different editions of each textbook

In order to do this, we will employ a similar analytic strategy as we did with gender, creating visualizations that reflect the racial composition of the artists. By crafting bar graphs that categorize artists by race, we can begin to identify patterns and measure the extent of racial diversity—or lack thereof—across the textbooks. Let's begin by making a bar graph of "artist_race" in the dataset:

```
# Visualize artist race distribution using a bar plot
artists |>
  ggplot(aes(x=artist_race)) +
  geom_bar() +
  coord_flip()
```

As seen in Fig. 8.4, the skew of white artists is particularly high across both of the fields. As we did with gender, we can review the representation of various racial categories across editions, and between Janson's *History of Art* and Gardner's *Art Through the Ages*. Again, we will remove the 1.1 edition on Janson's book for graphical clarity.

```
# Exclude edition 1.1 as it is not relevant for this analysis and add a
↪  bar plot with proportional representation by race
artists |>
  filter(edition_number != 1.1) |>
  ggplot(aes(x = edition_number, fill = artist_race)) +
  geom_bar(position = "fill") +
  facet_wrap(~book, scale = "free") +
  coord_flip()
```

Figure 8.5 reveals a cautiously optimistic signal: newer editions of both Janson's *History of Art* and Gardner's *Art Through the Ages* show an increased proportion of

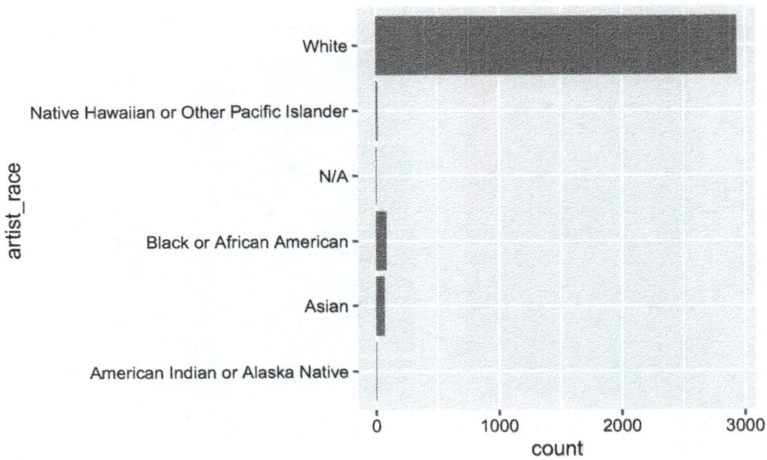

Fig. 8.4 Distribution of artist races in the art history textbook dataset

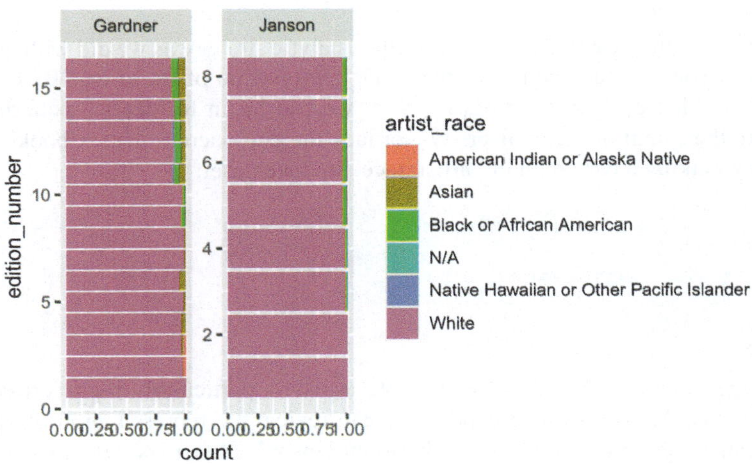

Fig. 8.5 Proportional representation of artist races across different editions of each textbook

non-white artists. This growth in representation, while a step in the right direction, falls short of capturing the full spectrum of racial diversity that exists within the art world. Much like the plateau observed in the representation of female artists, the advances in racial representation also appear to have hit a standstill. Furthermore, despite the overall trend toward increased inclusion, Janson's textbook appears to have made less progress in comparison to Gardner's, which raises important considerations about the editorial choices and the potential institutional biases that shape these educational resources.

The dataset curated we have does not merely catalog the presence of artists within the art history textbooks but also ventures into an analysis of the space each artist occupies—a metric that could reveal another layer of disparity or visibility. This aspect of the data, examining the ratio of page space devoted to each artist, invites a more nuanced exploration into the representation of artists. It is not just the frequency of occurrence that matters but the extent of the spotlight each artist receives. The space allocated to an artist in textbooks can influence the reader's perception of their significance and impact within the art world.

To interrogate this dimension, we will construct a histogram that quantifies the amount of space dedicated to artists in the textbooks. The histogram will provide a visual distribution of space allocation, revealing whether there is a general tendency to allocate more page space to certain artists and if this correlates with gender or other demographic factors. We will begin with a basic histogram that shows the space for all artists regardless of their demographics:

```
# Create a histogram to visualize the distribution of space ratio per
↪  page
artists |>
  ggplot(aes(x=space_ratio_per_page_total)) +
  geom_histogram()
```

Figure 8.6 shows that we have a very high right skew, showing that most artists tend to take up relatively small proportions of the page in each book. On the other hand, there is also a notable number of artists who are given a large amount of space on the page. This distribution of space given to artists could potentially have profound implications for how different groups of artists are perceived and evaluated

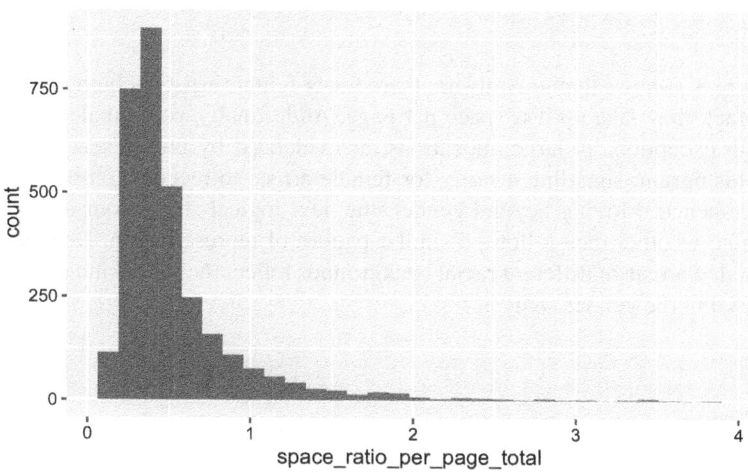

Fig. 8.6 Distribution of the space ratio allocated per page to artists

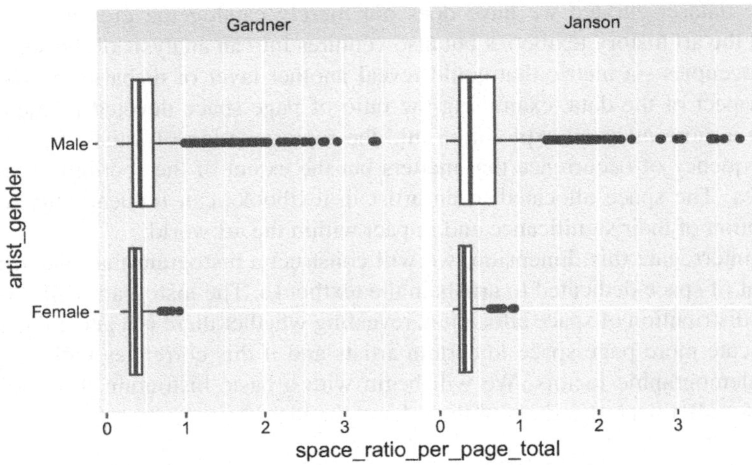

Fig. 8.7 Comparison of space ratio distribution by artist gender, faceted by textbook

in the art world. Let's see if female artists are more likely to have larger physical coverage than male artists in both books; we can do this using a boxplot.

```
# Create a plot with artist_gender on the x-axis and
↪  space_ratio_per_page_total on the y-axis. Add boxplots to visualize
↪  distribution
artists |>
  ggplot(aes(x=artist_gender, y=space_ratio_per_page_total)) +
  geom_boxplot() +
  facet_wrap(~book) +
  coord_flip()
```

Figure 8.7 shows that not only are there fewer female artists in both of the books, but in fact, they take up less space per page. Additionally, some male artists stand out with exceptionally large allocations, as evidenced by the presence of outliers in the histogram, signaling a rarity for female artists to receive comparable levels of prominence. Moving beyond gender, the next logical step in our analysis is to determine whether race follows a similar pattern of representation. Does the space allocated to artists of different racial backgrounds reflect the same kind of imbalance witnessed in the gender analysis?

```
# Create a plot with artist_race on the x-axis and
↪  space_ratio_per_page_total on the y-axis. Add boxplots to visualize
↪  distribution
artists |>
  ggplot(aes(x=artist_race, y=space_ratio_per_page_total)) +
  geom_boxplot() +
  facet_wrap(~book) +
  coord_flip()
```

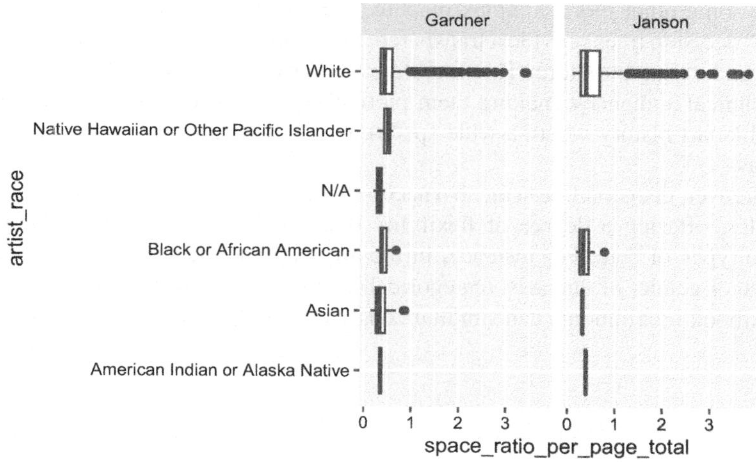

Fig. 8.8 Comparison of space ratio distribution by artist race, faceted by textbook

Unfortunately, Fig. 8.8 shows that there is little space dedicated to non-white artists. The underrepresentation, in many ways, perpetuates a biased and narrow understanding of art history, reinforcing existing power imbalances and excluding important perspectives. This not only limits the knowledge and understanding of students but also perpetuates systemic inequalities within the art world. Overall, our exploratory analysis shows us that both Gardner and Janson's textbooks are dominated by white male artists. However, what exactly is the relationship between the area of the page and race and gender? Throughout the rest of this chapter, we will use Tidymodel's implementation of decision trees to examine this relationship in more depth.

8.4 What Are Decision Trees?

In the context of cultural analytics, which often grapples with complex and nuanced patterns of human behavior, decision trees offer a powerful means of making sense of the data due to their ability to handle nonlinear and multifaceted relationships. They achieve this by iteratively partitioning the dataset into increasingly specific subsets based on the data's attributes. The resulting tree is composed of decision nodes and leaf nodes, each representing a crucial step in the decision-making process. Decision nodes act as junctures where the data is split according to a certain criterion related to a variable, such as the nationality or gender of an artist. Leaf nodes, on the other hand, represent the outcome of these decisions—the class label or decision reached after evaluating all the data's attributes (Loh 2014). Perhaps one of the chief benefits of decision trees is their interpretability.

Unlike some other more complex machine learning models that can act as "black boxes," decision trees provide a transparent and intuitive visual representation of the data's underlying structure. This facilitates easier communication of the results to nontechnical audiences, making them particularly well-suited for interdisciplinary fields like art history where a wide spectrum of scholars needs to engage with the findings.

Moreover, decision trees can also accommodate both categorical and continuous variables, offering a degree of flexibility that can be crucial when dealing with diverse types of data. For instance, in the art textbook analysis, variables such as the race or gender of the artist (categorical) and the space their artwork occupies in the textbook (continuous) can simultaneously inform the tree's construction.

8.5 Avoiding Overfitting

Many of the techniques that we have used so far have specifically focused on exploring or analyzing our dataset to understand certain phenomena. For instance, when looking at regression, we were able to fit a model to our data and determine coefficients that showed us the relationship between our dependent and independent variables. However, when working with machine learning algorithms, we will need to take a different approach. Rather than focusing on understanding and analyzing the data we have, we aim to develop models that can make predictions or decisions about new data. In other words, our focus shifts from gaining insights or explanations about the data to optimizing more accurate predictions.

A key element in predicting is to avoid overfitting the data, which occurs when a model becomes too complex and starts capturing noise or random variance in the data instead of the underlying patterns and relationships. This can lead to poor performance and inaccurate predictions of new or unseen data. To prevent overfitting, it is important to strike a balance between model complexity and generalization through techniques such as cross-validation, regularization, and feature selection.

For our purposes, we will focus on cross-validation. In this method, we start by creating subsamples of our data. We will use some of these samples to train our data while ignoring others. At first, this may seem counterintuitive: Why would we want to hold out data that could be used for training? The reason for doing so is that this provides us an estimate of how well the model generalizes to unseen data. If our model does incredibly well on our training data but poorly on the holdout data, we know we have overfit our model.

There are many forms of cross-validation. One of the most computationally expensive is leave-one-out cross-validation. In this form of cross-validation, the model is fit on all the data except one entry (e.g., an entry of a single artist in the art history books). You then fit this model a multitude of times with small adjustments made during each fit. While this allows us to maximize our data for training and testing purposes, the fact that it is time-consuming and resource-intensive means that it is rarely done. Instead, we can divide our data into a set-defined number of "folds." This is known as k-fold validation, with "k" representing the number of folds, with the most common values being k=5 or k=10.

Another more simple form of cross-validation is known as hold-out validation. In this, we divide our data into two sets: the training and testing set. Depending on the size of our data, the exact ratio for the two sets can vary. When we have a lot of data, even a small amount of our data might explain the variance, which means we may want 90% for the training set and only 10% for the testing set. For smaller datasets like ours, values, such as 75/25 and 80/20 are also common.

8.5.1 Using Resample to Split Data

The "resample" package, which comes with "tidymodels," has a handy `initial_split()` function that allows us to create our splits easily. By default, it will default to a 75/25 split, but we can specify a different proportion using the `prop` argument. Because our dataset is not very large, we will use a 70/30 split.

We will also use stratification, which can be specified using the `strata` argument. Stratification is useful when dealing with imbalanced data, such as what we see in the "space_ratio_per_page_total" column. When we created a histogram of this column, we saw that most artists were given little space. If we simply randomly split different observations, we may find that the split isn't representative of the entire dataset. To counter this, stratification ensures that the proportion of values is the same in all created sub-datasets as they were in the original dataset. For instance, if only 20% of the artists belong to a certain ethnic group, then this 20% would be reflected consistently across each sub-dataset. This helps maintain model accuracy and balance in all rounds of cross-validation, making sure no specific group's representation is artificially inflated or suppressed due to certain outliers or skewed training data.

Because the split is done randomly, we will also set the seed to "123" using the `set.seed()` function. This assures reproducibility and accuracy in our findings. Be sure to set the seed to the same value if you want to get the same results as shown in the rest of the chapter.

```
# Set the seed for reproducibility of random operations
set.seed(123)

# Split the data into training (70%) and testing (30%) sets using
↪  stratification
artists_split <- initial_split(artists, prop = 0.7, strata =
↪  space_ratio_per_page_total)

# Check the created split object
artists_split
```

```
<Training/Testing/Total>
<2172/932/3104>
```

The result of our split is an `rsplit` object that contains two different datasets. Approximately 70% of our total data (2212/3162 observations) has been assigned to the training set, while the remaining 30% (950/3162 observations) are in the testing set. Working with the `rsplit` object directly can be challenging. That's why "tidymodels" provides us with the `training()` and `testing()` functions to easily access this dataset and convert the `rsplit` object into two tibbles.

```
# Extract the training set from the split data
artists_training <- training(artists_split)

# Extract the testing set from the split data
artists_testing <- testing(artists_split)
```

8.6 Specifying and Running a Basic Model

Now that we have split our data into two tibbles, we need to specify our model structure. For this, we will use the "parsnip" package that is already installed when we load "tidymodels." With "parsnip," we have a consistent interface for specifying our model regardless of the package used. Because "tidymodels" draws inspiration from the tidyverse, we can use the pipe operator to specify the elements of our model, and at any point, we can save the result into an R object.

For a decision tree, the first step to declaring the model type is to simply use the `decision_tree()` function. It is important to note that this does not fit the model. At this stage, think of this primarily as an outline or blueprint of the decision tree model that you want to fit. This general approach is consistent with "tidymodels" philosophies of being compact and modular. With our `decision_tree()` function stored inside a variable, we can then further outline our model by adding model hyperparameters and choosing an engine (the package that actually performs the computation).

We will store the result of our call to the `decision_tree()` function in an R object we will name "artists_model."

```
# Initialize a decision tree model without specifying hyperparameters or
↪ data
artists_model <- decision_tree()

# Output the object to check the model's initial structure
artists_model
```

```
Decision Tree Model Specification (unknown mode)

Computational engine: rpart
```

We can see that the resulting object tells us we're creating a decision tree model, but it's still empty. It doesn't know where the data is coming from or what parameters to set for the model.

If we wanted to create a different type of model, we would only need to change this function. For instance, to create a simple linear regression, we can use the `linear_reg()` function. Likewise, for logistic regression, we can use the `logistic_reg()` function.

Some models, like logistic regression, can only be used (or should only be used) for classification. Decision trees, however, can be used for both regression (continuous variables) and classification (categorical variables) problems. Consequently, we need to specify the "mode." Because the ratio per page is a continuous variable, we will set our model "mode" to regression through the `set_mode()` function.

```
# Define a decision tree model using tidymodels' parsnip package
artists_model <- decision_tree() |>

# Set model mode for continuous outcomes (regression)
set_mode("regression")

# Display model specification summary
artists_model
```

```
Decision Tree Model Specification (regression)

Computational engine: rpart
```

We see that our model is now designated for regression. To complete the model specification, we need to specify the "engine"—the package that "tidymodels" will use to fit the decision tree model. In R, there are normally numerous packages that can be used to implement any type of machine learning model. One of the benefits of "parsnip" is that you can easily switch between different machine learning engines without changing the syntax of your code. Each engine might solve the same problem, but they all have different benefits and nuances, like computational efficiency, accuracy, and interpretability. You can find the list of engines for any model type by using the `show_engines()` function with a string indicating the type of model. The following shows how you would find the available "engines" for decision trees:

```
# Display available engines for decision tree models
show_engines("decision_tree")
```

```
# A tibble: 5 x 2
  engine mode
  <chr>  <chr>
1 rpart  classification
2 rpart  regression
3 C5.0   classification
```

```
4 spark   classification
5 spark   regression
```

As you can see, there are numerous engines for both classification and regression. We will use the "rpart" engine, which is arguably the most popular and well-established library for decision trees in R. To set the engine in our model "blueprint," we need to utilize the `set_engine()` function, feeding it with the string representing the desired engine ("rpart" in this case).

```
# Create a decision tree using parsnip. Set the model's mode to
↪ 'regression' for continuous outcomes and set the engine to 'rpart'
artists_model <- decision_tree() |>
  set_mode("regression") |>
  set_engine("rpart")

# Display model specification summary
artists_model
```

```
Decision Tree Model Specification (regression)

Computational engine: rpart
```

This is all that is needed to develop a model in "tidymodels." We can now "fit" the model on our training data. As we did with the `lm()` function for linear regression, we can use R's formula notation to choose our predictors and our outcome variables. For now, we will look at gender and race to space_ratio_per_page_total. Many people store the results of the fitted model in a separate object. Where exactly you distinguish a "model" and "fitting" depends on the context. I find it easier to think of fitting the model as part of the model itself. That way, this single object can be used in the future to make predictions. Because "parsnip" works with a single type of model object, you can make as many or as few "intermediary" objects as required.

```
# Fitting model using "gender" to predict "space ratio" on training data
artists_model <- decision_tree() |>
  set_mode("regression") |>
  set_engine("rpart") |>
  fit(space_ratio_per_page_total ~ artist_gender, data =
  ↪  artists_training)
```

To look at the results of our model, we only need to print the `artists_model` object:

```
# Display model specification summary
artists_model
```

```
parsnip model object

n= 2172

node), split, n, deviance, yval
      * denotes terminal node

1) root 2172 316.969800 0.5278105
  2) artist_gender=Female 242   4.332857 0.3685083 *
  3) artist_gender=Male 1930 305.725700 0.5477852 *
```

The output here is not very informative. It tells us that we have fit a decision tree model using "parsnip" on the artist training dataset, with a total of 2212 observations. The node depicts where the splits in the decision tree occur based on the predictor variables included in the model, and how many observations and deviance are associated with each node. For decision trees, it is usually easier to visualize the results of the model. The "rpart.plot" library contains a very handy function for doing this called `rpart.plot()`. This will produce a graphical display of the decision tree, with nodes that show the splits in the data based on the predictor variables. You will need to install and load the library if you have not already.

````
```{r}
# Install the rpart.plot package from CRAN
install.packages("rpart.plot")
# Load the rpart.plot library for use
library(rpart.plot)
```
````

Now that we have installed and loaded the package, we can produce a visualization of the decision tree created with our artists_model. However, we run into a problem: the "rpart.plot" package is not designed to work with "tidymodels" directly, which means we need to extract the fitted model object. We can do this using the `extract_fit_engine()` function. This function is useful for pulling out the underlying model object that "parsnip" has created to make it easier to work with other functions outside of "parsnip" that expect different kinds of model objects. Afterward, we can use the `rpart.plot()` function to visualize the decision tree. We will set `roundint` to `FALSE` to avoid rounding the labels in the plot.

```
# Use extract_fit_engine() to extract the underlying fitted model
extracted_model <- extract_fit_engine(artists_model)

# Plot the extracted decision tree model using rpart.plot()
rpart.plot(extracted_model, roundint = FALSE)
```

The simplicity shown in Fig. 8.9, indicated by its shallow depth and reliance on a single predictor—"artist-gender"—suggests that there is potential to refine and enhance the model to better understand the nuances of our data. The inclusion of additional predictors could illuminate other factors that influence the space ratio per page that each artist receives. Furthermore, we can explore the decision tree's

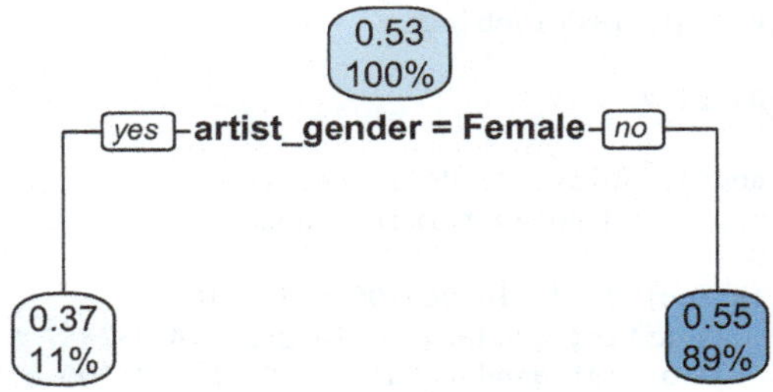

Fig. 8.9 Decision tree predicting space ratio based solely on artist gender

hyperparameters, which are the settings that govern the tree's construction process. By fine-tuning these hyperparameters, we might prevent overfitting, where the model becomes too tailored to the training data and loses its ability to generalize to new data.

For now, we will proceed with the current model by applying it to our testing data. This step is crucial for evaluating the model's predictive power and its ability to generalize beyond the data on which it was trained. By comparing the model's predictions against the actual space ratios in the testing set, we can measure its accuracy and identify areas where it may need improvement.

8.6.1 Checking Fit and Adding Additional Parameters

Now that we have trained our model on our training data and fit it, we can see how well it works on our testing data. To do this, we will use the `augment()` function. This function takes a fitted model and test data as inputs. It will then use the model rules established on the training dataset to predict response values for the test dataset. Finally, it will then bind two new columns entitled ".pred" and ".resid" to our original data.

```
# Use 'augment' to apply the model to the testing data and check results
artists_results <- augment(artists_model, artists_testing)

# Display the structure of the results for inspection
glimpse(artists_results)
```

```
Rows: 932
Columns: 16
$ .pred                      <dbl> 0.5477852, 0.5477852, 0.5477852, 0.5477852,~
$ .resid                     <dbl> -0.244525910, -0.094916671, -0.065521439, -~
$ artist_name                <chr> "Aaron Douglas", "Aaron Douglas", "Aaron Do~
$ edition_number             <dbl> 11, 14, 15, 16, 16, 5, 6, 8, 11, 12, 14, 4,~
$ year                       <dbl> 2001, 2013, 2016, 2020, 2020, 1970, 1975, 1~
$ artist_nationality         <chr> "American", "American", "American", "Americ~
$ artist_nationality_other   <chr> "American", "American", "American", "Americ~
$ artist_gender              <chr> "Male", "Male", "Male", "Male", "Female", "~
```

```
$ artist_race                 <chr> "Black or African American", "Black or Afri~
$ artist_ethnicity            <chr> "Not Hispanic or Latino origin", "Not Hispa~
$ book                        <chr> "Gardner", "Gardner", "Gardner", "Gardner",~
$ space_ratio_per_page_total <dbl> 0.3032593, 0.4528686, 0.4822638, 0.5437377,~
$ artist_unique_id            <dbl> 2, 2, 2, 2, 4, 6, 6, 6, 6, 7, 7, 8, 8, 15, ~
$ moma_count_to_year          <dbl> 0, 0, 0, 0, 0, 0, 0, 0, 0, 2, 2, 1, 1, 2, 4~
$ whitney_count_to_year       <dbl> 0, 0, 0, 0, 0, 0, 0, 0, 0, 0, 0, 0, 0, 0, 0~
$ artist_race_nwi             <chr> "Non-White", "Non-White", "Non-White", "Non~
```

We see that the `augment()` function has added the predicted values and the residuals for each observation in the testing set. It may be easier to select only the necessary data to see how well our model is working.

```
# Extract and examine predicted vs actual ratios for page space
artists_results |>
  select(space_ratio_per_page_total, .pred)
```

```
# A tibble: 932 x 2
   space_ratio_per_page_total .pred
                        <dbl> <dbl>
 1                      0.303 0.548
 2                      0.453 0.548
 3                      0.482 0.548
 4                      0.544 0.548
 5                      0.534 0.369
 6                      0.296 0.548
 7                      0.298 0.548
 8                      0.532 0.548
 9                      0.218 0.548
10                      0.457 0.548
# i 922 more rows
```

Based on a quick glance, it seems that our data is not providing us with very good results. One thing we can do is add additional predictors to our model. For instance, we can consider the number of exhibits an artist has had at the Museum of Modern Art (MoMA) or the Whitney Museum of American Art as a predictor. It is possible that as artists gain more exposure and have more exhibits in these prestigious art institutions, their page ratio in the textbooks may increase. By incorporating this information into our model, we may be able to improve its performance and accuracy. We can do this easily by simply adjusting our formula notation in the `fit()` function:

```
# Fit model with additional predictors from art museums
artists_model <- decision_tree() |>
  set_mode("regression") |>
  set_engine("rpart") |>
  fit(space_ratio_per_page_total ~ artist_gender + moma_count_to_year +
  ↪  whitney_count_to_year, data = artists_training)
```

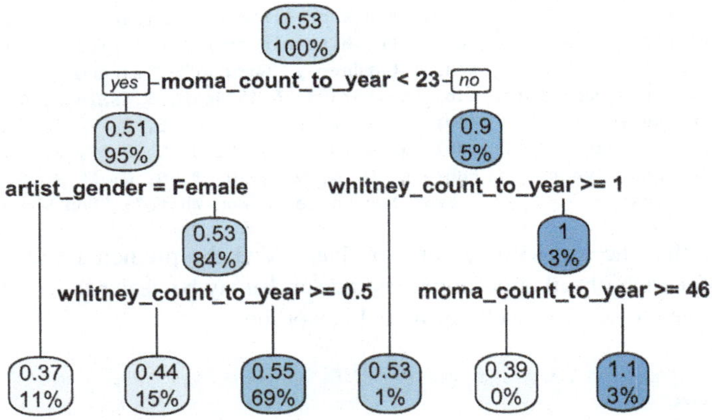

Fig. 8.10 Decision tree predicting space ratio based on gender and museum exhibition counts

Just as before, we would want to visualize the results of our updated model using the `rpart.plot()` function and get a better understanding of the splits in our decision tree.

```
# Extract fitted "rpart" model object from artists_model. Plot the
↪ decision tree
extracted_model <- extract_fit_engine(artists_model)
  rpart.plot(extracted_model)
```

As shown in Fig. 8.10, it is evident that after adding the exhibition count of the two art museums, the structure of our tree has changed significantly, indicating that the additional attributes have an impact on the outcome.

Let's now assess the quality of our updated model on the testing dataset. We can run the `augment()` function as before and filter the data we need out of it.

```
# Augment the model using testing set, predicting using decision tree
↪ model
artists_results <- augment(artists_model, artists_testing)

# Select and display predicted and actual space_ratio values from the
↪ results
artists_results |>
  select(.pred, space_ratio_per_page_total)
```

```
# A tibble: 932 x 2
   .pred space_ratio_per_page_total
   <dbl>                      <dbl>
 1 0.549                      0.303
 2 0.549                      0.453
 3 0.549                      0.482
 4 0.549                      0.544
```

```
 5 0.369                          0.534
 6 0.549                          0.296
 7 0.549                          0.298
 8 0.549                          0.532
 9 0.549                          0.218
10 0.549                          0.457
# i 922 more rows
```

By scanning the data, it appears that the model has improved slightly after adding our new predictors. However, how do we quantify this improvement? Luckily, "tidymodels" comes with a package called "yardstick" that we can use to quickly calculate measures. A common measurement for numeric predictors is the root mean squared error which calculates the square root of the mean squared differences between our predicted and actual values. A smaller RMSE indicates that the model is a good fit with a RMSE value of 0 showing that the model perfectly predicts our data.

To get the Root Mean Squared Value with "yardstick," we can use the `rmse()` function. This function takes a tibble or data frame as its first argument. In addition, it needs two additional arguments that represent the truth and the estimates. In our context, the "truth" column would be "space_ratio_per_page_total," while the estimates would be the ".pred" column.

```
# Calculate RMSE comparing truth vs predictions
artists_results |>
  rmse(truth = space_ratio_per_page_total, estimate = .pred)
```

```
# A tibble: 1 x 3
  .metric .estimator .estimate
  <chr>   <chr>          <dbl>
1 rmse    standard       0.409
```

We get a result of approximately 0.41. Overall, these results are a good start, but there is room for further improvement.

8.6.2 Adding Hyperparameters

Hyperparameters are additional parameters that can fine-tune our decision tree model. Although there are numerous ones we can control, we will focus on cost complexity and tree depth. Cost complexity controls the size of the decision tree by creating trade-offs between following each branch and making additional splits. It seeks to balance the increase in model complexity for an equivalent reduction in error. In other words, the model avoids splitting branches into subbranches if the error reduction it receives is below a threshold. When we declare the model type as `decision_tree()`, it uses a default cost complexity value. We can, however, tweak this according to our needs through the `cost_complexity` parameter. Let's set the parameter to 0.1 to see how well our model performs under these conditions.

```
# Create a decision tree model with cost complexity. Fit the model to
↪ training data with specified predictors
artists_model <- decision_tree(cost_complexity = 0.1) |>
  set_mode("regression") |>
  set_engine("rpart") |>
  fit(space_ratio_per_page_total ~ artist_gender + moma_count_to_year +
  ↪ whitney_count_to_year, data = artists_training)
```

Let's check the `rmse()` to see if it has improved.

```
# Bind predicted values to the test data
artists_testing |>
  bind_cols(predict(artists_model, artists_testing)) |>
  rmse(truth = space_ratio_per_page_total, estimate = .pred)
```

```
# A tibble: 1 x 3
  .metric .estimator .estimate
  <chr>   <chr>          <dbl>
1 rmse    standard       0.430
```

Upon evaluating the refined decision tree model, which now includes more predictors and adjusted hyperparameters, we observe a marginal enhancement in its performance. While any improvement is welcome, the incremental gains suggest we are approaching a threshold beyond which additional tweaks to this particular model will yield minimal benefits. In the face of such a plateau, our analysis would be well-served by considering alternative approaches.

8.7 What Are Random Forest Models?

One of the benefits of "tidymodels" is that we can easily change our model definition and have it evaluate different machine learning methods. Having trimmed and refined our decision tree model, let's now work with a more complex and robust method: random forests (Breiman 2001).

Random forest models are a powerful and versatile method for making predictions. They consist of an ensemble or collection of decision trees, where each tree is trained on a different subset of the data. The predictions of the individual trees are then combined to make a final prediction. This helps to avoid overfitting, because by training multiple trees on different subsets of the data, we can reduce the impact of individual noisy or irrelevant features and capture the more robust patterns in the data.

Another benefit of random forest models is their improved prediction accuracy. By aggregating the predictions of multiple trees, the model can account for different sources of variability in the data and arrive at a more reliable prediction. Any errors or biases that may occur in individual decision trees are smoothed out across the ensemble, resulting in a more robust and accurate overall performance. This allows them to be generally less sensitive to the specific choices of hyperparameters compared to solitary decision trees and achieve good performance without as much fine-tuning.

These advantages have resulted in random forests being widely used in a variety of domains. In the field of cultural analytics, random forests can provide valuable insights into patterns and trends within artistic and cultural data. To understand how to incorporate random forests into our work, we will take a look at workflows in "tidymodels."

8.8 Workflows

Workflows in the "tidymodels" framework are designed to simplify and enhance the model-building experience. They provide a coherent and structured approach to assembling the various components required for machine learning tasks. Workflows encapsulate the entire machine learning process within a single object, enabling users to avoid repetitive coding and potential errors that can occur when these steps are handled separately. A typical workflow in "tidymodels" will consist of at least two critical components: the model specification and a recipe. The model specification, which we have already seen, outlines the type of model to be built, including the algorithm and its associated modes and engines. Recipes, which we will look at later, are akin to a blueprint for data preprocessing. It includes all the preprocessing instructions to be applied to the dataset before training the model. These instructions might involve normalization or scaling of variables, encoding of categorical features, imputation of missing values, and even the creation of new features through various transformations. The recipe ensures that data preprocessing is reproducible and consistent, which is vital for the integrity of the model's performance.

To start a workflow, we simply call the `workflow()` function and save the results into an object.

Here's how:

```
# Initialize a new workflow object
artists_workflow <- workflow()

# Display the empty workflow object
artists_workflow
```

```
== Workflow ===================================================================
Preprocessor: None
Model: None
```

As you can see, this object does not contain anything. We will need to add both our model and our recipe to it. Let's begin with the model specification. We saw how to create a model specification for decision trees, and the steps to create a random forest specification are quite similar. We only need to set our `engine()` and `mode()` to be appropriate for random forests. Since we are using the "ranger" package, you may need to install and load it if you do not already have it.

```
# Uncomment to install ranger if necessary
# install.packages("ranger")
library(ranger)

# Define classification mode for random forest using "ranger" package as
↪ engine
artists_model <- rand_forest() |>
  set_mode("classification") |>
  set_engine("ranger")
```

Once the model specification is established, integrating it into the workflow is the next pivotal step. This integration is achieved using the `add_model()` function, which is a key aspect of the "tidymodels" workflow construction process. By employing this function, the specified model becomes a component of the larger workflow object.

```
# Add the defined random forest model to the workflow
artists_workflow <- artists_workflow |>
  add_model(artists_model)
```

Let's take a look at what we have so far:

```
# Examine the workflow with the model added
artists_workflow
```

```
== Workflow ====================================================================
Preprocessor: None
Model: rand_forest()

-- Model -----------------------------------------------------------------------
Random Forest Model Specification (classification)

Computational engine: ranger
```

8.8.1 Recipes

So far, we have implemented our model and manually adjusted it. However, for the sake of convenience and automation, "tidymodels" provides a package called "recipes," which we can harness to quickly conduct data preprocessing and feature engineering. These steps are often necessary before feeding the data into our models and can take up a large chunk of the overall model-creation process. Different models need different types of preprocessed data. As such, a very common practice is to carry out several preprocessing tasks such as centering and scaling numeric variables, encoding categorical variables, handling missing data, or even creating interaction variables.

Depending on the complexity of the data and the specific model requirements, the preprocessing steps can become quite complicated, time-consuming, and prone to errors. Luckily, random forests require little or no preprocessing of data, as they can handle both categorical and numerical variables and are largely robust to outliers. However, to get a foundation for preprocessing you may need in the future, let's take a look at how we can use the "recipes" package in "tidymodels."

The first step in creating a recipe is to declare the formula that we are interested in predicting. We will use all of the data that we have minus the artist_name.

```
# Creating a recipe for preprocessing the training data. Specifying
↪ prediction formula and identifying training data
artists_recipe <-
recipe(artist_gender ~ edition_number + year +
        space_ratio_per_page_total + moma_count_to_year +
        whitney_count_to_year + artist_race_nwi,
        data = artists_training)
```

Once we have written the formula for the recipe, we can add steps that describe the transformations we want to perform on the data. Following the tidyverse syntax, these added steps typically start with the term `step_`, followed by an action for the transformation. The parameterization of each `step_` function is highly intuitive; it directly references the columns of the dataset on which the transformation should be applied. For example, if we want to normalize numeric predictors, the `step_normalize()` function can be called with a selector that identifies these numeric columns. This selector could be as broad as `all_numeric()` for all numeric columns, or it could be more finely tuned to target specific variables.

We can also impute values with recipes. For instance, the `step_impute_median()` function is a straightforward approach where each missing value in a numeric column is replaced with the median of the available values in that column. This method is robust to outliers and can be a sensible choice when the data distribution is skewed. By strategically adding these imputation steps into a recipe, we can systematically address missing data and improve the robustness of the subsequent model.

In our case, we will implement two preprocessing steps. One, we will use the `step_dummy()` function to create dummy, or indicator, variables for each categorical variable, with the exception of our target outcome variable. Two, we will focus on standardizing the numerical variables through the `step_normalize()` function. This step will transform our numeric predictors, so that they're centered around a mean of zero and scaled to a standard deviation of one. Such normalization is a critical process, especially when different scales across variables could otherwise introduce a bias in the learning algorithm.

Let's see how to add these steps to our recipe:

```
# Establishes base recipe with formula for predicting artist gender.
↪ Expand nominal vars into dummy vars, excludes outcome (target)
↪ variable. Normalize all numeric vars to have mean = 0, sd = 1
artists_recipe <- artists_recipe |>
  step_dummy(all_nominal(), -all_outcomes()) |>
  step_normalize(all_numeric())
```

Having created our recipe and added all the necessary preprocessing steps, we can then integrate our recipe into the previously created workflow using the `add_recipe()` function as follows. Let's go ahead and do this and look at the results:

```
# Add the recipe object to the workflow for preprocessing steps
artists_workflow <- artists_workflow |>
  add_recipe(artists_recipe)

# Display the workflow to inspect model and recipe details
artists_workflow
```

```
== Workflow ====================================================================
Preprocessor: Recipe
Model: rand_forest()

-- Preprocessor ----------------------------------------------------------------
2 Recipe Steps

* step_dummy()
* step_normalize()

-- Model -----------------------------------------------------------------------
Random Forest Model Specification (classification)

Computational engine: ranger
```

The workflow we've constructed now encapsulates the full scope of our data processing and modeling strategy. It combines the model—the engine of our predictions, with the recipe, our carefully thought-out plan for data preparation. This unified object can now be used to "fit" our data as before.

8.8.2 Fitting Random Forest Models with Workflows

With our random forest model specification and preprocessing recipe firmly established within our workflow, we can fit our random forest model on the training data.

```
# Fit the random forest model using the preprocessed training data
artists_fit <-
  artists_workflow |>
  fit(data = artists_training)
```

The output of our fit is a **workflow_fit** object containing our preprocessed training data, the learned model parameters, and the model metrics. We can now see how well this data predicts gender just as before using the **augment()** function.

```
# Use augment to apply the fitted model to the testing data
artists_results <- augment(artists_fit, new_data = artists_testing)

# Display the structure of the results for inspection
glimpse(artists_results)
```

```
Rows: 932
Columns: 17
$ .pred_class                <fct> Male, Male, Male, Male, Male, Male, Male, M~
$ .pred_Female               <dbl> 0.090920956, 0.108921195, 0.297988076, 0.39~
$ .pred_Male                 <dbl> 0.9090790, 0.8910788, 0.7020119, 0.6020960,~
$ artist_name                <chr> "Aaron Douglas", "Aaron Douglas", "Aaron Do~
$ edition_number             <dbl> 11, 14, 15, 16, 16, 5, 6, 8, 11, 12, 14, 4,~
$ year                       <dbl> 2001, 2013, 2016, 2020, 2020, 1970, 1975, 1~
$ artist_nationality         <chr> "American", "American", "American", "Americ~
$ artist_nationality_other   <chr> "American", "American", "American", "Americ~
$ artist_gender              <chr> "Male", "Male", "Male", "Male", "Female", "~
$ artist_race                <chr> "Black or African American", "Black or Afri~
$ artist_ethnicity           <chr> "Not Hispanic or Latino origin", "Not Hispa~
$ book                       <chr> "Gardner", "Gardner", "Gardner", "Gardner",~
$ space_ratio_per_page_total <dbl> 0.3032593, 0.4528686, 0.4822638, 0.5437377,~
$ artist_unique_id           <dbl> 2, 2, 2, 2, 4, 6, 6, 6, 6, 7, 7, 8, 8, 15, ~
$ moma_count_to_year         <dbl> 0, 0, 0, 0, 0, 0, 0, 0, 0, 2, 2, 1, 1, 2, 4~
$ whitney_count_to_year      <dbl> 0, 0, 0, 0, 0, 0, 0, 0, 0, 0, 0, 0, 0, 0, 0~
$ artist_race_nwi            <chr> "Non-White", "Non-White", "Non-White", "Non~
```

Remember, it's always a good practice to evaluate your model after it's been initially trained. While we have previously discussed using the Root Mean Squared Error (RMSE) to assess model performance for continuous data, this metric falls short when it comes to categorical values such as gender. In these instances, we turn to the confusion matrix, a powerful tool that elucidates the accuracy of classification models by revealing the nature of prediction errors. A confusion matrix lays out the outcomes of our model's predictions in four quadrants. On one side, we have the counts for when the model rightly identifies an artist's gender, giving us instances where the predicted gender aligns with the true gender—these are our accurate hits. On the flip side, we see when the model misses the mark, predicting one gender when the artist is actually identified as another.

Constructing a confusion matrix with the yardstick package requires both the actual gender classifications from our test data and the corresponding predictions made by the model. These inputs must be factor variables, which R uses to

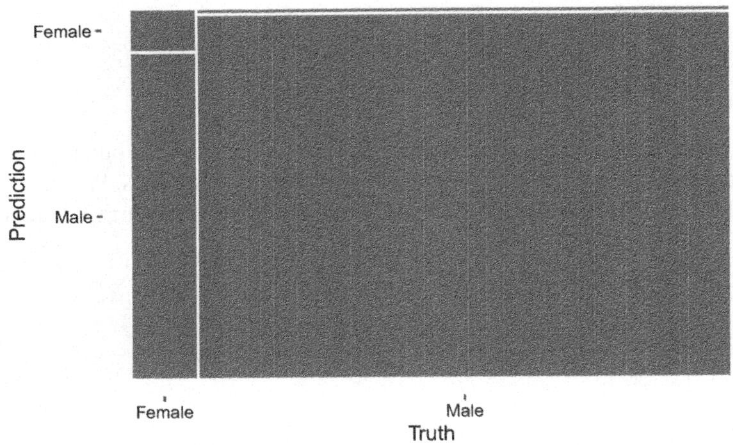

Fig. 8.11 Confusion matrix visualizing the performance of the random forest model predicting artist gender

handle categorical data with proper levels and classifications. So, we will need to convert any character-based gender representation to factors with appropriate levels representing the different possible genders. Once we have converted our gender variables to factors, we can pipe the results into the `confusion-matrix()` function which takes two additional inputs: the truth and the estimate. Rather than just printing the confusion matrix as a table, we can pipe the results into the `autoplot()` function enabling us to create a visualized confusion matrix which can be much easier to interpret and assess visually, as shown in Fig. 8.11.

```
# Convert character vector to factor for proper confusion matrix
↪  generation
artists_results$artist_gender <- as.factor(artists_results$artist_gender)

# Generate confusion matrix using predictions and true gender labels
artists_results |>
  conf_mat(truth = artist_gender, estimate = .pred_class) |>
  autoplot()
```

From these results, we can observe the model's performance characteristics. While the overall accuracy is relatively high at approximately 89.8% (calculated as (11+826)/932), this masks significant class imbalance issues. The model is much better at identifying "Male" artists (True Negatives = 826) than "Female" artists (True Positives = 11). The high number of False Positives (89) indicates that many "Female" artists were incorrectly classified as "Male," contributing to a very low precision of 11% for the "Female" class (11 / (11+89)). However, the low number of False Negatives (6) results in a higher recall of approximately 64.7% for the "Female" class (11 / (11+6)), meaning that when an artist *is* female, the model correctly identifies them about two-thirds of the time. Despite this recall rate, the model rarely predicts "Female." This imbalance suggests potential issues like uneven training data or features biased toward "Male" characteristics, necessitating improvements such as data rebalancing or feature reevaluation.

8.9 Tuning Hyperparameters with Grids

We saw that our random forests model worked relatively well with the default settings. However, for any form of predictive modeling, you should always look for ways to better increase the accuracy or precision of your model. Earlier, we saw how to adjust hyperparameters with our decision tree by simply selecting a different hyperparameter, `cost_complexity`, that governed the complexity of the decision tree in the `decision_tree()` function. But, how do we know if the hyperparameters that we have chosen are optimal for our model? The answer to this problem is hyperparameter tuning.

Hyperparameter tuning is the process of selecting the most suitable set of hyperparameters for our model. There are various ways this can be done, but for our purpose, we will use a method called "grid search." Grid search allows you to specify various combinations of hyperparameters that you wish to test with each combination of parameters used to fit a mode. The performance is then assessed via cross-validation, and the set of hyperparameters that yield the best models is chosen as the optimal set for the random forest model.

8.9.1 The Three Parts of a Grid

Creating a comprehensive grid can be a resource-intensive task, particularly for models with high complexity or when attempting to explore a broad range of hyperparameter values. The computational cost increases with the size of the grid and the complexity of the model being tuned, as each unique set of hyperparameters requires the model to be trained and evaluated. As such, finding the right balance between exhaustiveness and practicality is key.

To actually create a grid, we will use the `tune_grid()` function. This function is designed to streamline hyperparameter exploration by integrating seamlessly with the modeling workflow. The three critical components required by `tune_grid()` are:

1. Workflow: This encapsulates the entire modeling process, including both the preprocessing steps and the model specification. The workflow ensures that the same sequence of operations is applied to the data for each set of hyperparameters tested.

2. Cross-validation set: To reliably assess model performance, we employ cross-validation, which involves partitioning the dataset into subsets. The model is then trained on some of these subsets and validated on others, reducing the risk of overfitting to the training data. This cross-validation set is pivotal in the tuning process, as it allows the `tune_grid()` function to evaluate the efficacy of each hyperparameter combination across different slices of the data.
3. Grid specification: This is the hyperparameter grid outlining the range and combinations of hyperparameter values to be tested. The grid specification guides the tuning process by providing the `tune_grid()` function with the distinct configurations to apply to the model.

Let's go ahead and create each of these in turn. We will begin by specifying the model for our workflow just as before. However, this time, we will specify that we want to tune three hyperparameters: `mtry`, `min_n`, and `tree_depth`. Together, these control the number of variables considered at each split, the minimal node size, and the maximum depth of any tree in the random forest, respectively. To do this, we will simply set the value of the parameters to `tune()` in our model specification.

```
# Define random forest model with tuned tree count and node size

artists_rf_tune <-
  rand_forest(mtry = tune(), trees = tune(), min_n = tune()) |>
  set_engine("ranger") |>
  set_mode("classification")

# Initialize workflow. Add model specification to to workflow and  add
↪ data preprocessing recipe to workflow
artists_workflow_tune <- workflow() |>
  add_model(artists_rf_tune) |>
  add_recipe(artists_recipe)
```

Now that we have established a workflow, our next step is to prepare a cross-validation set. To enhance model validation, rather than solely relying on a single training and testing set division, we can employ k-fold cross-validation. If you recall from earlier, k-fold cross-validation is a systematic method that involves partitioning the data into "k" distinct segments, or "folds." In this process, the model is trained on (k-1) of these folds and then tested on the remaining fold. With each iteration, a different fold is used for validation, allowing the model to be trained and validated across all segments of the data. This procedure is repeated until each fold has been used once as the validation set. By averaging the prediction errors across all k iterations, we obtain a more comprehensive and reliable assessment of the model's performance.

To implement k-fold cross validation, we will utilize the `vfold_cv()` and specify "10" as the number of folds to generate from our training data. Additionally, we will apply stratification by setting the `strata` argument to "artist_gender." This ensures that each fold contains a proportional representation of genders, reflecting the balance present in the entire dataset, and supports the maintenance of model integrity and relevance across diverse subsets of data.

```
# Perform 10-fold cross-validation on training data and stratify folds by
↪ gender
artists_folds <- vfold_cv(artists_training, v = 10, strata =
↪ artist_gender)
```

Once our folds are created, we can go to the final step of specifying a grid. For our example, we will create a Latin hypercube using the `grid_space_filling()` function. This advanced grid generation technique requires us to provide the set of hyperparameters we intend to explore, which we extract from our random forest model specification using `extract_parameter_set_dials()`. However, some hyperparameters, like `mtry` (the number of predictors to sample at each split), depend on the actual data. Therefore, before generating the grid, we must use the `finalize()` function, providing it with our training data (`artists_training`). This step allows `dials` to determine the appropriate range for `mtry` based on the number of predictors present in the training set. After finalizing, we can proceed with `grid_space_filling()`.

The Latin hypercube sampling method approach enables us to cover a vast hyperparameter space with a manageable number of samples (McKay et al. 1979). By selecting a "size" parameter—let's say 30—we instruct the function to produce 30 unique, strategically distributed points within our hypercube. Think of this as creating a representative "sampling" of the vast hyperparameter space, which allows us to bypass the need for exhaustive permutation testing and instead focus on a curated subset that promises a high likelihood of containing the optimal set. This method not only accelerates the tuning process but also ensures that the resulting grid is both diverse and representative, increasing the chances of pinpointing a near-optimal combination without the prohibitive computational load typically associated with such extensive searches.

```
#Generate hyperparameter grid using Latin hypercube sampling method, size
↪ of 30 and extract tunable parameters from random forest model; use
↪ for grid setup
artists_grid <- artists_rf_tune |>
   extract_parameter_set_dials() |>
   finalize(artists_training) |>
    grid_space_filling(size = 30)
```

Let's take a look at what our tuning grid looks like so far:

```
# Display the tuning grid to inspect the hyperparameter combinations
artists_grid
```

```
# A tibble: 30 x 3
    mtry trees min_n
   <int> <int> <int>
```

```
1      1   621    28
2      1  1310    19
3      1   414    11
4      2  1379    33
5      2  1103     5
6      2  1793     9
7      3  1931    25
8      3   138    29
9      4   759    38
10     4   690    20
# i 20 more rows
```

As we can see, the tuning grid consists of 30 unique combinations of the three hyperparameters we identified previously: `mtry`, `min_n`, and `trees`.

We can now finally tune our grid. To do this, we use the `tune_grid()` function which takes the name of the workflow, the grid object, and the resamples we created through k-fold cross-validation as arguments. We will also use the `control_grid()` function to save our workflow, which will come in handy later when we want to revert and manually inspect our models.

```
# Apply grid search using tuned workflow on the folds for hyperparameter
↪  tuning
tune_results <- tune_grid(
  artists_workflow_tune,
  resamples = artists_folds,
  grid = artists_grid,
  control = control_grid(save_workflow = TRUE)
)
```

After tuning, Fig. 8.12 visualizes our tuning results using "tidymodels" plotting functions, laying out the performances for each combination. The code to do this is as follows:

```
# autoplot function visualizes tuning results across hyperparameter
↪  values
autoplot(tune_results)
```

We have now completed the hyperparameter tuning process. We can inspect our results to identify the top-performing model from the grid search and use it to predict new data, leading to potentially greater accuracy with our random forest model. If you prefer to see the results by themselves, you can also get the results of the model by using the `collect_metrics()` function.

```
# Store all performance metrics from tuning in 'final_results'
final_results <- collect_metrics(tune_results)
```

This function will return a tibble, i.e., a tidyverse data frame, that contains the metrics for each model we have considered in the grid search. Each row corresponds to a different model, and each column corresponds to a different metric or hyperparameter. We can get the best model by using the `show_best()` function.

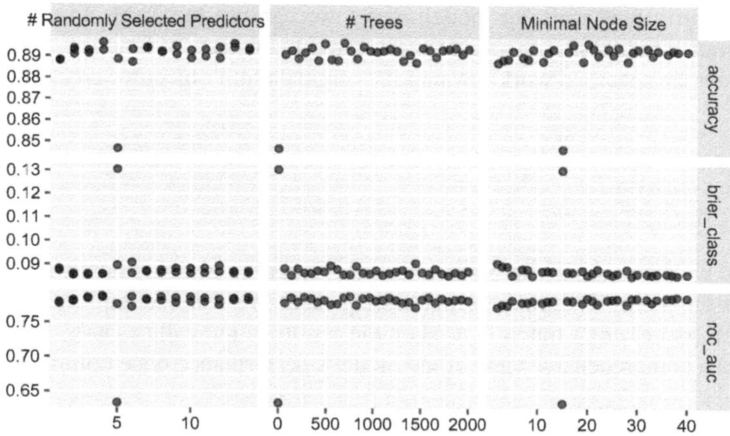

Fig. 8.12 Visualization of hyperparameter tuning results for the random forest model

```
# Displays top models from tuning based on accuracy
show_best(tune_results)
```

```
# A tibble: 5 x 9
   mtry trees min_n .metric .estimator  mean     n std_err .config
  <int> <int> <int> <chr>   <chr>      <dbl> <int>   <dbl> <chr>
1     4   759    38 roc_auc binary     0.789    10  0.0121 Preprocessor1_Model09
2     8   276    36 roc_auc binary     0.787    10  0.0131 Preprocessor1_Model17
3    10   966    40 roc_auc binary     0.787    10  0.0135 Preprocessor1_Model22
4     6  1586    37 roc_auc binary     0.787    10  0.0130 Preprocessor1_Model13
5     4   690    20 roc_auc binary     0.787    10  0.0135 Preprocessor1_Model10
```

This function provides a table representation showing the best performing model(s) concerning the accuracy and other measure levels, optimizing the process of choosing the superior one accordingly. We can save this data using the `select_best()` function. This will automatically train the model with the best set of hyperparameters on the training data.

```
# Fit the model with optimal hyperparameters from tuning results
best_param <- select_best(tune_results)
```

Finally, we can finalize our workflow by using the `finalize_workflow()` and sending the best parameters as the argument. This will ensure that the model is trained with the best hyperparameters.

```
# Finalize the model for fitting
artists_workflow_tune <- artists_workflow_tune |>
  finalize_workflow(best_param)
```

To quantify how well our model works, we can use the `last_fit()` function to generate predictions on our test set. The `last_fit()` function takes as its arguments the specified workflow and the original rsplit object we created.

```
# Apply the best model from tuning to the last training and validation
↪ set
final_predictions <-
  last_fit(artists_workflow_tune, artists_split)
```

Once we've obtained the predictions, it's critical to evaluate how well they line up with the actual outcomes in our testing data. We can accomplish this via the `collect_metrics()` function used on the results of our call to `last_fit()`. This provides a comprehensive view of the model's performance once optimized.

```
# Collect model performance metrics from the final prediction results
collect_metrics(final_predictions)
```

```
# A tibble: 3 x 4
  .metric     .estimator .estimate .config
  <chr>       <chr>          <dbl> <chr>
1 accuracy    binary        0.899  Preprocessor1_Model1
2 roc_auc     binary        0.804  Preprocessor1_Model1
3 brier_class binary        0.0823 Preprocessor1_Model1
```

It returns metrics such as accuracy, roc_auc (for classification tasks), rmse (for regression tasks), and others that help us understand the model's effectiveness.

8.10 Conclusion

Throughout this chapter, we ventured into the intricacies of model optimization with a focus on improving our understanding of a cultural analytics dataset. We utilized the "tidymodels" library in R to fine-tune a machine learning model, specifically applying a Latin hypercube-generated grid to examine a multitude of hyperparameter combinations. The process was not only methodical but also insightful, leading us to uncover an optimal set of hyperparameters that substantially enhanced our model's predictive accuracy.

This meticulous optimization effort yielded a model that provides us with a clearer lens through which to view our cultural data. As a case in point, we were able to delve deeper into the representation patterns of artists in art history textbooks—a topic of cultural significance—and unearth biases in gender and ethnicity representation. By adjusting the hyperparameters, we improved the model's ability to discern subtle patterns and relationships within the data, which in turn allowed us to draw more nuanced conclusions about the prevalence and depiction of different demographic groups within influential educational materials.

Overall, we have seen how the “tidymodels” framework is a powerful asset for scholars in cultural analytics, offering a comprehensive suite of tools designed for machine learning and statistical modeling within the R programming environment. While all the steps may be challenging for newcomers to the field, the principles and practices we applied are universally valuable across various domains of data science. As new techniques emerge and the field continues to evolve, knowing the foundations will allow you to remain adaptable to these changes and continue contributing meaningful insights drawn from complex data.

Chapter 9
Conclusion

Abstract The final chapter of this book emphasizes the boundless nature of cultural analytics, driven by vast datasets and innovative methodologies. It introduces three emerging areas of research—spatial analysis, advanced machine learning, and visual data analysis—that offer new horizons for scholars and practitioners. This section serves as an inspirational springboard, urging readers to extend their knowledge and engage in the dynamic field of cultural analytics beyond the book's scope.

Keywords Cultural analytics · Spatial analysis · Advanced machine learning · Geographic information systems (GIS) · Neural networks · Computational image analysis · Deep learning

In the pursuit of mastering R for the specific application of cultural analytics, this book has meticulously traversed the multifaceted landscape of data analysis and visualization, extending into the realms of machine learning and network science. By this point, readers have not only been introduced to the R programming language but have also been equipped with a robust toolkit that harmonizes the principles of tidy data with the intricacies of cultural datasets.

As we reach the final pages of our journey together, it is important to acknowledge that the world of cultural analytics is boundless, with limitless potential for exploration and discovery. The techniques and insights presented throughout this book provide a sturdy launchpad for delving into the vast universe of data-driven cultural research. Yet, myriad avenues within this field remain untraversed.

9.1 Brief Foray into Further Explorations

In the forthcoming sections, we will focus on three promising domains that hold particular intrigue for those seeking to expand their cultural analytics toolkit beyond the scope of this book. These areas—spatial analysis, advanced machine learning techniques, and visual analysis—represent the frontiers of current research

N. Siddiqui, *Cultural Analytics in R: A Tidy Approach*, Use R!,
https://doi.org/10.1007/978-3-031-96618-7_9

and application in cultural analytics. Each offers a different lens through which to interpret the vast tapestry of cultural data, providing fresh perspectives and challenges that invite you to apply, adapt, and extend the skills you have cultivated. As with all exploratory endeavors, they also present the chance to contribute to the ongoing dialogue and development of the field, pushing the boundaries of what is possible with cultural data. The goal is not to provide an exhaustive treatise on these subjects but to inspire you to continue exploring and learning.

9.1.1 Spatial Analysis

In the digital humanities and computational social sciences, the utility of mapping and spatial analysis is often not fully realized, overshadowed by the prominence of textual and numerical data analysis. Yet, the integration of spatial information into cultural analytics is a powerful catalyst for discovery, illuminating the physical context in which cultural phenomena occur. Geographic information systems (GIS) are pivotal in this integration, equipping cultural analysts with the ability not only to visualize but also to interrogate the geographical dimensions of their data.

The "sf" package in R, which integrates seamlessly with the tidyverse, exemplifies the harmony between spatial data and the tidy data principles that have been a recurring theme throughout this book (Pebesma et al., 2023). "sf" simplifies handling spatial data by providing a framework that is both intuitive and consistent with the grammar and syntax of tidy data manipulation. Whether dealing with points, lines, or polygons, "sf" enables cultural analysts to work with spatial objects like any other tidy dataset. This means that spatial data can be manipulated, transformed, and visualized using familiar tidyverse tools, maintaining a fluent and efficient workflow. By harnessing the capabilities of "sf" alongside the tidy data paradigm, analysts can uncover the hidden spatial relationships within cultural data and weave these findings into the larger tapestry of cultural analytics, adding depth and dimension to their insights.

9.1.2 Advanced Machine Learning Techniques

We have already seen how "tidymodels" can allow us to create robust predictive models, especially with techniques such as random forests and decision trees. These models have opened a window into the predictive possibilities within cultural datasets, providing us with powerful tools to discern patterns, classify cultural artifacts, and forecast future trends based on historical data. The decision tree method, with its clear logic and interpretability, serves as an excellent starting point for those new to machine learning, while random forests enhance predictive accuracy through ensemble learning, reducing the risk of overfitting inherent to single decision trees.

However, the landscape of machine learning within R stretches far beyond these methods, encompassing a variety of advanced techniques that can further enhance the cultural analyst's toolkit. Techniques such as gradient boosting machines (GBM), support vector machines (SVM), and neural networks represent more advanced forms of algorithmic prediction and classification. Each of these techniques brings a unique approach to model building, from GBM's iterative refinement of models to SVM's optimization for decision boundaries. By incorporating these advanced machine learning techniques, cultural analysts can tackle increasingly sophisticated questions and datasets, uncovering nuanced patterns and associations that may be imperceptible through simpler models.

9.1.3 Visual Data

In the digital humanities, text remains the dominant medium of expression and analysis (Manovich, 2020; Sayers, 2018; Meeks, 2013). This focus on text has led some to characterize the field as disproportionately skewed toward textual analysis, trailing in both data visualization and computational simulations (Champion, 2017). Nonetheless, the burgeoning availability of visual data combined with advancements in computational image analysis heralds a transformative shift in scholarship possibilities. Wevers and Smits ascribe this paradigmatic shift, often referred to as the "visual turn," to the development and dispersion of sophisticated computer vision algorithms (Wevers and Smits, 2020). These algorithms are particularly effective when harnessing the power of deep neural networks for image interpretation. These networks are adept at detecting and synthesizing hierarchical structures within images, applying layers of specialized filters to discern contours, textures, and essential visual features. This methodical dissection of visual data engenders a richer and progressively intricate understanding of the content—an innovative frontier in digital humanities research.

Unfortunately, R is not typically seen as the first choice for implementing deep learning models geared toward visual data analysis, as languages like Python with libraries such as TensorFlow and PyTorch have dominated this space. Yet, with the right packages and interfaces, R too can be leveraged for sophisticated image analysis tasks. The "keras" package in R, for instance, provides a comprehensive high-level neural networks API that can access TensorFlow functionality, thereby facilitating deep learning endeavors within the R environment (Kalinowski et al., 2023). In addition, "imager" is another R package that offers a set of image processing capabilities for analyzing and manipulating visual data (Barthelme et al., 2023). Using "imager," analysts can perform operations such as cropping, rotating, filtering, and even segmenting images, enabling them to prepare and refine visual datasets for subsequent machine learning analyses.

9.2 What's Next?

As we close this book, remember that the journey with R and cultural analytics does not end here. The landscapes of culture and data are ever-evolving, and there will always be new questions to ask, new datasets to explore, and new insights to unearth. The skills and knowledge you have acquired are stepping stones to further discovery, critical analysis, and innovative research in the rich, interdisciplinary field of cultural analytics.

This book is not merely a conclusion but a commencement—a launchpad into the exciting world of data-driven cultural exploration. With the command of R and an appreciation for tidy data, you are now equipped to embark on your own analytical adventures, to uncover the stories hidden within the numbers, and to contribute meaningfully to the evolving dialogue of culture in the digital age.

References

About Gapminder | Gapminder. n.d. https://www.gapminder.org/about/about-gapminder/.

Akiki, Christopher, and Manuel Burghardt. 2021. MuSe: The Musical Sentiment Dataset. *Journal of Open Humanities Data* 7. https://openhumanitiesdata.metajnl.com/articles/10.5334/johd.33.

Álvarez, Adolfo. 2021. Plumbers, Chains, and Famous Painters: The (Updated) History of the Pipe Operator in R, September. http://adolfoalvarez.cl/blog/2021-09-16-plumbers-chains-and-famous-painters-the-history-of-the-pipe-operator-in-r/.

Arnold, Jeffrey. 2024. *Jrnold/Ggthemes*. https://github.com/jrnold/ggthemes.

Arnold, Taylor, Ken Benoit, Lincoln Mullen, Adam Obeng, and rOpenSci Text Workshop Participants (2017). 2023. *Tif: Text Interchange Format*. https://docs.ropensci.org/tif.

Arnold, Taylor, and Lauren Tilton. 2015. *Humanities Data in r: Exploring Networks, Geospatial Data, Images, and Text*. Quantitative Methods in the Humanities and Social Sciences. Cham: Springer International Publishing.

Arnold, Taylor, and Lauren Tilton. 2019. *New Data? The Role of Statistics in Digital Humanities*. Minneapolis, MN: University of Minnesota Press.

Bache, Stefan Milton, Hadley Wickham, and Lionel Henry. 2022. *Magrittr: A Forward-Pipe Operator for r*. https://cloud.r-project.org/web/packages/magrittr/index.html.

Barthelme, Simon, David Tschumperle, Jan Wijffels, Haz Edine Assemlal, Shota Ochi, Aaron Robotham, and Rodrigo Tobar. 2023. *Imager: Image Processing Library Based on 'CImg'*. https://cran.r-project.org/web/packages/imager/index.html.

Benjamin, Daniel J., and James O. Berger. 2019. Three Recommendations for Improving the Use of p-Values. *The American Statistician* 73 (1): 186–91.

Bertin, Jacques. 1983. *Semiology of Graphics: Diagrams, Networks, Maps*. Madison, WI: University of Wisconsin Press.

Birjali, Marouane, Mohammed Kasri, and Abderrahim Beni-Hssane. 2021. A Comprehensive Survey on Sentiment Analysis: Approaches, Challenges and Trends. *Knowledge-Based Systems* 226: 107134.

Blondel, Vincent D., Jean-Loup Guillaume, Renaud Lambiotte, and Etienne Lefebvre. 2008. Fast Unfolding of Communities in Large Networks. *Journal of Statistical Mechanics: Theory and Experiment* 2008 (10): P10008. https://doi.org/10.1088/1742-5468/2008/10/P10008.

Bowser, Pearl, Jane Gaines, and Charles Musser, eds. 2016. *Oscar Micheaux & His Circle: African-American Filmmaking and Race Cinema of the Silent Era*. Bloomington, IN: Indiana University Press.

Bowser, Pearl, and Louise Spence. 2000. *Writing Himself into History: Oscar Micheaux, His Silent Films, and His Audiences*. New Brunswick, NJ: Rutgers University Press.

N. Siddiqui, *Cultural Analytics in R: A Tidy Approach*, Use R!,
https://doi.org/10.1007/978-3-031-96618-7

Breiman, Leo. 2001. Random Forests. *Machine Learning* 45 (1): 5–32. https://doi.org/10.1023/A:1010933404324.

Brown, Samuel. 1872. Report on the Eighth International Statistical Congress, Held at St. Petersburg, 22nd/10th August to 29th/17th, 1872. *Journal of the Statistical Society of London* 35 (4): 431–57. https://doi.org/10.2307/2338764.

Challenge, T7-Hackathon. n.d. Pokemon- Weedle's Cave. https://www.kaggle.com/datasets/terminus7/pokemon-challenge.

Champion, Erik Malcolm. 2017. Digital Humanities Is Text Heavy, Visualization Light, and Simulation Poor. *Digital Scholarship in the Humanities* 32: i25–32.

Cifor, Marika, Hanna Girma, Shanya Norman, and Miriam Posner. 2018a. Definition | Early African Film Database. https://earlyracefilm.github.io/database/definition.html.

Cifor, Marika, Hanna Girma, Shanya Norman, and Miriam Posner. 2018b. Early African-American Film Database, 1909-1930 4 (0): 1. https://doi.org/10.5334/johd.7.

Codd, Edgar F. 1990. *The Relational Model for Database Management: Version 2*. Reading, MA: Addison-Wesley.

Cripps, Thomas. 1979. *Black film as genre*. Midland books. Bloomington, IN: Indiana University Press.

Csardi, Gabor, and Tamas Nepusz. 2006. The Igraph Software Package for Complex Network Research. *InterJournal, Complex Systems* 1695 (5): 19. https://www.researchgate.net/profile/Gabor-Csardi/publication/221995787_The_Igraph_Software_Package_for_Complex_Network_Research/links/0c96051d301a30f265000000/The-Igraph-Software-Package-for-Complex-Network-Research.pdf.

Das, Kousik, Sovan Samanta, and Madhumangal Pal. 2018. Study on Centrality Measures in Social Networks: A Survey. *Social Network Analysis and Mining* 8 (13). https://doi.org/10.1007/s13278-018-0493-2.

Dennis, Tim, and Doug Daniels. n.d. Gapminder. https://doi.org/10.7910/DVN/GJQNEQ.

Desagulier, Guillaume. 2017. *Corpus Linguistics and Statistics with r: Introduction to Quantitative Methods in Linguistics*. Quantitative Methods in the Humanities and Social Sciences. Cham: Springer International Publishing. https://doi.org/10.1007/978-3-319-64572-8.

Diawara, Manthia, ed. 1993. *Black American Cinema*. AFI Film Readers. New York: Routledge. http://www.gbv.de/dms/bowker/toc/9780415903974.pdf.

Fruchterman, Thomas M. J., and Edward M. Reingold. 1991. Graph Drawing by Force-Directed Placement. *Software: Practice and Experience* 21 (11): 1129–64. https://doi.org/10.1002/spe.4380211102.

Gaines, Jane. 2001. *Fire and desire: Mixed-race movies in the silent era*. Chicago, IL: University of Chicago Press. http://hdl.handle.net/2027/heb.05931.

Ghosh, Soumitra, and Brojo Kishore Mishra. 2020. Parts-of-Speech Tagging in Nlp: Utility, Types, and Some Popular Pos Taggers, 131–65. Palm Bay, FL: CRC Press.

Greenland, Sander. 2019. Valid p-Values Behave Exactly as They Should: Some Misleading Criticisms of p-Values and Their Resolution with s-Values. *The American Statistician* 73: 106–14.

Griffin, Randi. 2018. Kagglers Love My Olympic History Dataset! https://www.randigriffin.com/2018/08/17/kagglers-love-my-data.html.

Grün, Bettina, Kurt Hornik, David M. Blei, John D. Lafferty, Xuan-Hieu Phan, Makoto Matsumoto, Takuji Nishimura, and Shawn Cokus. 2023. *Topicmodels: Topic Models*. https://cran.r-project.org/web/packages/topicmodels/index.html.

HaCohen-Kerner, Yaakov, Daniel Miller, and Yair Yigal. 2020. The Influence of Preprocessing on Text Classification Using a Bag-of-Words Representation. *PloS One* 15 (5).

Harper, F. Maxwell, and Joseph A. Konstan. 2015. The MovieLens Datasets: History and Context. *ACM Transactions on Interactive Intelligent Systems* 5 (4): 19:1–19. https://doi.org/10.1145/2827872.

Ihaka, Ross. 1998. R: Past and Future History. *Computing Science and Statistics*. https://www.stat.auckland.ac.nz/~ihaka/downloads/Interface98.pdf.

Jelodar, Hamed, Yongli Wang, Chi Yuan, Xia Feng, Xiahui Jiang, Yanchao Li, and Liang Zhao. 2019. Latent Dirichlet Allocation (LDA) and Topic Modeling: Models, Applications, a Survey. *Multimedia Tools and Applications* 78 (11): 15169–211. https://doi.org/10.1007/s11042-018-6894-4.

Jockers, Matthew L., and Rosamond Thalken. 2020. *Text Analysis with R: For Students of Literature*. Quantitative Methods in the Humanities and Social Sciences. Cham, Switzerland: Springer International Publishing. https://doi.org/10.1007/978-3-030-39643-5.

Kalinowski, Tomasz, Daniel Falbel, J. J. Allaire, François Chollet, RStudio, Google, Yuan Tang [ctb, et al. 2023. *Keras: R Interface to 'Keras'*. https://cran.r-project.org/web/packages/keras/index.html.

Kuhn, Max. 2008. Building Predictive Models in r Using the Caret Package. *Journal of Statistical Software* 28 (5): 1–26. https://doi.org/10.18637/jss.v028.i05.

Kuhn, Max, and Hadley Wickham. 2020. *Tidymodels: A Collection of Packages for Modeling and Machine Learning Using Tidyverse Principles*. https://www.tidymodels.org.

Loh, Wei-Yin. 2014. Fifty Years of Classification and Regression Trees. *International Statistical Review* 82 (3): 329–48. https://doi.org/10.1111/insr.12016.

Lüdecke, Daniel, Mattan S. Ben-Shachar, Indrajeet Patil, Brenton M. Wiernik, Etienne Bacher, Rémi Thériault, and Dominique Makowski. 2022. Easystats: Framework for Easy Statistical Modeling, Visualization, and Reporting. *CRAN*. https://easystats.github.io/easystats/.

Manovich, Lev. 2016. The Science of Culture? Social Computing, Digital Humanities and Cultural Analytics. *Journal of Cultural Analytics* 1 (1). https://doi.org/10.22148/16.004.

Manovich, Lev. 2020. *Cultural Analytics*. Cambridge, MA: MIT Press.

Matloff, Norm. 2023. *Teaching r in a Kinder, Gentler, More Effective Manner:* https://github.com/matloff/TidyverseSkeptic.

McKay, M. D., R. J. Beckman, and W. J. Conover. 1979. A Comparison of Three Methods for Selecting Values of Input Variables in the Analysis of Output from a Computer Code. *Technometrics* 21 (2): 239–45. https://doi.org/10.2307/1268522.

Meeks, Elijah. 2013. Is Digital Humanities Too Text-Heavy? | Digital Humanities Specialist. https://dhs.stanford.edu/spatial-humanities/is-digital-humanities-too-text-heavy/.

Milligan, Ian. 2016. The Problem of History in the Age of Abundance, December. https://uwspace.uwaterloo.ca/handle/10012/11817.

Montoyo, Andrés, Patricio Martínez-Barco, and Alexandra Balahur. 2012. Subjectivity and Sentiment Analysis: An Overview of the Current State of the Area and Envisaged Developments. *Decision Support Systems* 53 (4): 675–79. https://www.sciencedirect.com/science/article/pii/S0167923612001339?casa_token=HyezXuEaivMAAAAA:DbFmCOrmBGf7vUZQD2E0i1iFP6IYntehZhOpVjKYZJYjWhbR3VRzM0oxnil7ggRNfOKjsjl-.

Müller, Kirill, and Hadley Wickham. 2023. *Tibble: Simple Data Frames*. https://github.com/tidyverse/tibble.

Nelder, John Ashworth, and Robert W. M. Wedderburn. 1972. Generalized Linear Models. *Journal of the Royal Statistical Society* 135 (3): 370–84.

Nielsen, Finn Årup. 2011. A New ANEW: Evaluation of a Word List for Sentiment Analysis in Microblogs. arXiv. https://doi.org/10.48550/arXiv.1103.2903.

Norman Matloff. 2014. R Beats Python! R Beats Julia! Anyone Else Wanna Challenge R? Mad (Data) Scientist. May 21, 2014. https://matloff.wordpress.com/2014/05/21/r-beats-python-r-beats-julia-anyone-else-wanna-challenge-r/.

Osborne, Jason, and Elaine Waters. 2019. Four Assumptions of Multiple Regression That Researchers Should Always Test. *Practical Assessment, Research, and Evaluation* 8 (1).

Palsky, Gilles. 1999. The Debate on the Standardization of Statistical Maps and Diagrams (1857–1901). Elements of the History of Graphical Semiotics. *Cybergeo: European Journal of Geography*, March. https://doi.org/10.4000/cybergeo.148.

Pebesma, Edzer, Roger Bivand, Etienne Racine, Michael Sumner, Ian Cook, Tim Keitt, Robin Lovelace, et al. 2023. *Sf: Simple Features for r*. https://cran.r-project.org/web/packages/sf/index.html.

Pedersen, Thomas Lin. 2017. Data Imaginist—Introducing Tidygraph. https://www.data-imaginist.com/posts/2017-07-07-Introducing-tidygraph/index.html.

Pedersen, Thomas Lin. 2022. Ggraph: An Implementation of Grammar of Graphics for Graphs and Networks.

Pedersen, Thomas Lin. 2023. *Tidygraph: A Tidy API for Graph Manipulation*. https://tidygraph.data-imaginist.com.

Posner, Miriam, and Matthew Lincoln. 2016. Cytoscape Tutorials, June. https://doi.org/10.5281/zenodo.56245.

Queiroz, Gabriela De, Colin Fay, Emil Hvitfeldt, Os Keyes, Kanishka Misra, Tim Mastny, Jeff Erickson, David Robinson, and Julia Silge. 2023. *Tidytext: Text Mining Using 'Dplyr', 'Ggplot2', and Other Tidy Tools*. https://cran.r-project.org/web/packages/tidytext/index.html.

Rattenbury, Tye, Joseph M. Hellerstein, Jeffrey Heer, Sean Kandel, and Connor Carreras. 2017. *Principles of Data Wrangling: Practical Techniques for Data Preparation*. Sebastopol, CA: O'Reilly Media.

Redfern, Nick. 2023. *Computational Film Analysis with r*. https://cfa-with-r.netlify.app/cfa.

Robertson, Stephen. 2004. Understanding Inverse Document Frequency: On Theoretical Arguments for IDF. *Journal of Documentation* 60 (5): 503–20. https://doi.org/10.1108/00220410410560582.

Rosenzweig, Roy. 2003. Scarcity or Abundance? Preserving the Past in a Digital Era. *The American Historical Review* 108 (3): 735–62. https://academic.oup.com/ahr/article-abstract/108/3/735/22504.

Sayers, Jentery. 2018. Introduction: Studying Media Through New Media, 1–6. New York, NY: Routledge.

Schöch, Christof. 2013. Big? Smart? Clean? Messy? Data in the Humanities. *Journal of Digital Humanities* 2 (3): 2–13.

Silge, Julia, and David Robinson. 2017. *Text Mining with r: A Tidy Approach*. 1st ed. Beijing, China; Boston, MA: O'Reilly.

Software Studies Initiative. 2015. Cultural Analytics. In Software Studies Initiative. 2015. http://lab.softwarestudies.com/p/overview-slides-and-video-articles-why.html.

Sparck Jones, Karen. 1972. A Statistical Interpretation of Term Specificity and Its Application in Retrieval. *Journal of Documentation* 28 (1): 11–21. https://doi.org/10.1108/eb026526.

Stam, Holland. n.d. Data from: Quantifying Art Historical Narratives. https://doi.org/10.7924/r4dn48h0w.

Statistical Analysis of Network Data with R. 2020. 2nd ed. Use R! Cham, Switzerland: Springer.

Thieme, Nick. 2018. R Generation. *Significance* 15 (4): 14–19. https://doi.org/10.1111/j.1740-9713.2018.01169.x.

Underwood, Ted. 2019. *Distant Horizons: Digital Evidence and Literary Change*. Chicago, IL: University of Chicago Press.

Uysal, Alper Kursat, and Serkan Gunal. 2014. The Impact of Preprocessing on Text Classification. *Information Processing & Management* 50 (1): 104–12.

Ward, Grady. 2006. *Moby Part of Speech List*. http://archive.org/details/mobypartofspeech03203gut.

Wellek, Stefan. 2017. A Critical Evaluation of the Current "p-Value Controversy". *Biometrical Journal* 59 (5): 854–72.

Wevers, Melvin, and Thomas Smits. 2020. The Visual Digital Turn: Using Neural Networks to Study Historical Images. *Digital Scholarship in the Humanities* 35 (1): 194–207.

Why Can't Ggplot2 Use. 2018. https://community.rstudio.com/t/why-cant-ggplot2-use/4372.

Wickham, Hadley. 2010a. A Layered Grammar of Graphics. *Journal of Computational and Graphical Statistics* 19 (1): 3–28. https://doi.org/10.1198/jcgs.2009.07098.

Wickham, Hadley. 2010b. A Layered Grammar of Graphics. *Journal of Computational and Graphical Statistics* 19 (1): 3–28. https://doi.org/10.1198/jcgs.2009.07098.

Wickham, Hadley. 2014. Tidy Data. *Journal of Statistical Software* 59 (September): 1–23. https://doi.org/10.18637/jss.v059.i10.

Wickham, Hadley. 2016. *Ggplot2: Elegant Graphics for Data Analysis*. New York, NY: Springer.

Wickham, Hadley. n.d. *The Tidyverse Style Guide*. https://style.tidyverse.org/index.html.

Wickham, Hadley, Mara Averick, Jennifer Bryan, Winston Chang, Lucy McGowan, Romain François, Garrett Grolemund, et al. 2019. Welcome to the Tidyverse. *Journal of Open Source Software* 4 (43): 1686. https://doi.org/10.21105/joss.01686.

Wickham, Hadley, Mine Çetinkaya-Rundel, and Garrett Grolemund. 2023. *R for Data Science: Import, Tidy, Transform, Visualize, and Model Data*. 2nd ed. Sebastopol, CA: O'Reilly.

Wilkinson, Leland. 2012. The Grammar of Graphics, ed. James E. Gentle, Wolfgang Karl Härdle, and Yuichi Mori, 375–414. Berlin, Germany: Springer. https://doi.org/10.1007/978-3-642-21551-3_13.

Wilkinson, Leland, and Graham Wills. 2005. *The Grammar of Graphics*. 2nd ed. In Statistics and Computing. New York: Springer.

Xie, Yihui. 2018. Knitr: A Comprehensive Tool for Reproducible Research in r, 3–31. Boca Raton, FL: Chapman; Hall/CRC.

Zhou, Tao, Jie Ren, Matúš Medo, and Yi-Cheng Zhang. 2007. Bipartite Network Projection and Personal Recommendation. *Physical Review E* 76 (4): 046115. https://doi.org/10.1103/PhysRevE.76.046115.

Zimmerman, Naupaka, Greg Wilson, Raniere Silva, Scott Ritchie, François Michonneau, Jeffrey Oliver, Harriet Dashnow, et al. 2019. Swcarpentry/r-Novice-Gapminder: Software Carpentry: R for Reproducible Scientific Analysis, June 2019. https://doi.org/10.5281/zenodo.3265164.

The manufacturer's authorised representative in the EU is Springer Nature Customer Service Centre GmbH, Europaplatz 3, 69115 Heidelberg, Germany. If you have any concerns regarding our products, please contact ProductSafety@springernature.com

Printed and bound by CPI Group (UK) Ltd, Croydon, CR0 4YY
30/09/2026
02225561-0006